Dolls and Accessories
of the 1930s and 1940s

Dian Zillner

Photography
by Suzanne Silverthorn

4880 Lower Valley Road, Atglen, PA 19310 USA

Introduction

The writing of this book has been a special treat for me. I have been able to relive my own childhood by remembering and picturing many of my dolls from the 1930s and 1940s. Most of my birthday and Christmas celebrations included the gift of a new doll and I received my last childhood doll at the age of twelve. My doll family was also expanded with the occasional purchase of an inexpensive composition or cloth doll with money I had saved to pay for the item. One of my 8" all composition toddler dolls was even named the Quarter Doll, reflecting her small cost. My mother made clothes for all my dolls and the Quarter Doll had a very large wardrobe. Since I kept only my "best" dolls, alas, my "Quarter Doll" and her extensive wardrobe were discarded many years ago.

It is surprising, after fifty to seventy years, how many lovely composition dolls still survive. Some of these dolls even retain their original tags. Since the composition material can easily crack or craze, it is very remarkable to find lovely examples, from a variety of companies, that have retained their beauty. These dolls, of course, bring top dollar in today's market. Most collectors will accept *some* craze lines in the composition dolls they purchase because that may be the only flaw on an otherwise beautiful doll. The dolls that are badly cracked, crazed with paint flecking, redressed, or repainted, however, do not sell well, even at reduced prices.

The top companies producing composition dolls during the 1930s and 1940s were the Alexander Doll Co., Effanbee Doll Corp., Ideal Toy Co., and the E.I. Horsman firm. For the most part, the Alexander and the Effanbee dolls are the most expensive to purchase. Besides these well-known firms, other companies also left their mark on the era. Included were Arranbee, American Character, Monica, and Eugenia.

Another interesting phenomenon that occurred during this period was the emergence of many people who have since become well-known in the doll collecting field because of their doll artistry or their new ideas in dolls marketing. These include Joseph Kallas, Dewees Cochran, Mary Hoyer, Georgene Hendren and, of course, Beatrice (Madame) Alexander.

Cloth dolls have been included in this book as well because these dolls were also popular with consumers in the 1930s and 1940s. Catalogs from the two decades picture a wide variety of cloth dolls, including those produced by the major doll companies: Alexander, Effanbee, and Ideal. In addition, the Boudoir doll was still popular during the 1930s and 1940s. These dolls were marketed in both cloth and composition materials. The painted over bisque Nancy Ann Storybook Dolls also rate a special place in a book on 1930s and 1940s dolls. These dolls were quite inexpensive when new, but have now become quite collectible with the rare dolls selling for hundreds of dollars.

An introduction to a book on 1930s and 1940s dolls cannot be complete without a few words about the many celebrity dolls sold during the era. Of course, Ideal's Shirley Temple dwarfed all other such dolls with her popularity. But there were many success stories as well for other celebrity dolls, including Deanna Durbin, Judy Garland, Charlie McCarthy, Sonja Henie, Jane Withers, Margaret O'Brien, and Princess Elizabeth. Other, lesser known celebrity dolls from the era are also covered in this book.

The price of a doll depends on the condition, popularity, and rarity of the doll. The prices in this book reflect the doll as it is pictured. If clothing has been replaced, this information will be included in the caption and the price given will be for a doll with replaced clothing. If the doll pictured includes the box, the price noted will reflect that extra addition. No price is listed for dolls pictured only in catalog advertisements in order not to confuse the reader about the condition of the doll (box or no box, tag or no tag, etc.). Some of the photographs in this book have been supplied by collectors who are very knowledgeable in their own field of expertise—Alexander, Nancy Ann Storybook, etc. In this way we hope to show the reader which of these various dolls are the most unusual and therefore would be priced higher than like dolls dressed in more common costumes.

I wish to thank the many collectors who have been so helpful in sharing materials, photographs, dolls, and knowledge with me so that this book could become a reality. These include Judith Armitstead, Jo Barckley, Linda Boltrek, Ellen Cahill, Dorothy Cassidy, Cobbs's Doll Auction, Frashers' Doll Auctions, Linda Friend at Friendly Antiques, Jan Hershey, Lois and Bob Jakubowski, Veronica Jochens, Bonnie McCullough, Betty Nichols, Marilyn Pittman, Gayle Reilly, Nancy Roeder, Carol Stover, Susie's Museum of Childhood, Mary Lu Trowbridge, and Phyllis Young. A special vote of thanks goes to Marge Meisinger who, once again, loaned me catalogs from her fabulous collection so advertisements from the era could be photographed. She also supplied pictures of dolls from her fantastic collection.

I would also like to extend continued appreciation and thanks to my daughter, Suzanne Silverthorn, who again took many of the photographs for this, my tenth book. A special "thank you" is also given to my son, Jeff Zillner, who helped with editing once again.

Acknowledgement and recognition is also extended to Schiffer Publishing Ltd., and its excellent staff, particularly to designer Sue Taylor and editor Donna Baker, who helped with this publication. Without their support and extra effort, this book would not have been possible.

Marilyn Pittman, aged nine, pictured with her childhood dolls. Marilyn kept many of her dolls and several are pictured in this book.

The author at the age of ten, pictured with one of two dolls she received as Christmas gifts in 1942. This is an unidentified cloth doll while the other doll was an Alexander baby doll with magic skin.

Alexander Doll Company

The Alexander Doll Company was founded by Beatrice Alexander Behrman in 1923. At first the dolls were made of cloth and the operation was mostly an "at home" kind of enterprise. Beatrice came into the doll business naturally as her father had opened the first doll hospital in the United States. She and her sisters were born over the shop in New York City.

After graduating from high school, Beatrice married Phillip Behrman in 1912. As her interest in doll making grew, she persuaded her husband to join her in the business. She began by producing cloth dolls with flat faces but soon molded faces were added to these dolls (see Cloth Dolls chapter).

By the late 1920s or early 1930s, it is likely that the New York Alexander firm was buying composition dolls from other companies, dressing them, and selling them under their own label. It is not known if Alexander manufactured the earliest Betty dolls (Mama dolls with cloth bodies) or if they bought them from another firm and then dressed and marketed them.

The company evidently did produce the 7" all composition Tiny Betty, the 9" all composition Little Betty, and the 13" all composition Betty dolls. These dolls, dressed in different costumes, were part of the Alexander line for many years.

The 7" Tiny Betty dolls had a one-piece body and head with movable arms and legs, painted features, mohair wigs, and molded shoes and socks. This model was used in producing a variety of Alexander dolls. The lines included those based on fairy tales, nursery rhymes, literary characters, "Dolls of the Month," and native costumes from countries around the world. The Little Women dolls were particularly popular. They were based on characters from Louisa Alcott's famous book of the same name.

The 9" all composition Little Betty dolls were fully jointed and also had painted features and mohair wigs. These dolls wore shoes and socks. They, too, were sold dressed in various costumes and were a part of the Alexander line for many years. These dolls were produced to represent characters from fairy tales, nursery rhymes, Little Women, "Days of the Week," McGuffey Ana, Little Colonel, and foreign countries. Both the Tiny Betty and Little Betty dolls were also dressed in many more different outfits including bridal attire.

The 13" all composition Betty dolls were fully jointed and had sleep eyes, closed mouths, and wigs. The dolls were dressed as "Betty" in little girl outfits, but were also used for a variety of "name" dolls, including Princess Elizabeth, the Quintuplet nurse, Little Colonel, and McGuffey Ana.

24" Alexander "My Betty," circa late 1920s-30. She has a composition head-shoulder plate, composition arms and legs, and cloth body. The doll has sleep eyes, open mouth, and a mohair wig. She is all original with her tag and box. The tag reads "My/Betty/MADAME ALEXANDER NEW YORK." Her dress is tagged "Madame Alexander/NEW YORK." It is very hard to find these early Alexander dolls with tags and the original box (not enough examples to determine a price). *Doll and photograph from the collection of Veronica Jochens.*

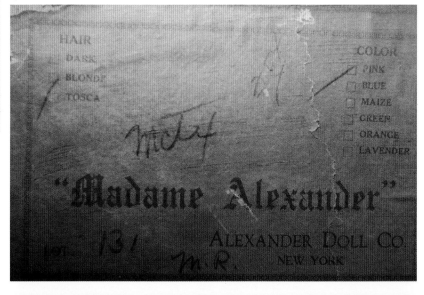

The "My Betty" box includes the marked color of dress and wig as well as Lot #131. The printing reads "Madame Alexander/ALEXANDER DOLL CO./NEW YORK." *Box and photograph from the collection of Veronica Jochens.*

13" composition Alexander Betty, circa 1935. She has sleep eyes, closed mouth, a mohair wig, and is fully jointed. The doll is unmarked and the dress appears to be original ($250-300). *Doll from the collection of Marge Meisinger. Photograph by Carol Stover.*

7" Alexander composition Tiny Betty dressed as a Scotch girl. She has painted eyes, closed mouth, mohair wig, and is jointed only at the shoulders and hips. She is all original complete with her original wrist tag. It reads "Created By/ MADAME ALEXANDER/New York N.Y./U.S.A." ($175 with tag). *Doll from the collection of Marge Meisinger. Photograph by Carol Stover.*

7" Alexander composition Tiny Betty dressed as Little Red Riding Hood. She has painted eyes, closed mouth, mohair wig, and is jointed only at the shoulders and hips. Her shoes and socks are molded. She is all original. These dolls, in a variety of outfits, were marketed by Alexander for many years. This doll appears to be circa mid to late 1930s ($225+). *Doll from the collection of Dana Gaddy. Photograph by Carol Stover.*

7" Alexander composition Tiny Betty Bo Peep, circa late 1930s to early 1940s. She is all original with a "Bo Peep" gold sticker on her slip ($250+). *Doll and photograph from the collection of Nancy Roeder.*

7" Alexander composition Tiny Betty Spanish doll MIB, circa late 1930s-early 1940s ($250+). *Doll and photograph from the collection of Nancy Roeder.*

By 1933, the firm was evidently producing their own dolls; in her book, *Collector's Encyclopedia of American Composition Dolls*, Ursula R. Mertz pictured an Alexander ad from 1933 promoting Walt Disney's "Three Little Pigs" dolls. She also showed an Alexander ad for "Betty's Troussew Carry-All Case" from 1934. The small composition doll in the set had molded hair similar to the Effanbee Patsy dolls. The kit included clothes to sew. Both ads appeared in *Playthings* magazines.

The Madame Alexander Company must have been selling unmarked composition dolls during the early 1930s as their dolls from this period are hard to identify. The firm was well enough established by 1935 to receive exclusive licensing to produce the Dionne Quintuplet dolls. Another personality doll, Baby Jane, was also marketed that year (all shown in Personality Dolls chapter).

In order to compete with the Ideal company and their Shirley Temple dolls, the Alexander firm secured the rights to produce "Little Colonel" dolls based on the leading character from the books written by Anne Fellows Johnston. These dolls were on the market by Christmas of 1935 to tie in to the Shirley Temple Fox film *Little Colonel*. The dolls came in sizes of 13", 16", 17", 22", and 23" in the Sears Christmas catalog of 1935. They all had open mouths except for the 13" size. Their wigs were mohair. The copy notes that the "clothing was designed by Madame Alexander — a leading doll stylist."

By 1937, the Alexander company was producing a variety of Dionne Quint dolls in both baby and little girl styles. They had also added another celebrity doll when they marketed the Jane Withers composition doll that year (see Personality Dolls chapter).

The 1937 Sears Christmas catalog pictured what appears to be a doll called "Special Girl" by collectors. The catalog referred to her as "Pigtails." She came in 16" and 24" sizes and had a composition head, arms and legs, a cloth body, real hair wig, closed mouth, and sleep eyes.

By 1938, the John Plain catalog was featuring the Alexander Princess Elizabeth dolls, which came in a variety of sizes. The small 13" doll was really the closed mouth Betty doll dressed as Princess Elizabeth. The larger dolls were made from their own Princess Elizabeth design, which included an open mouth with teeth and sleep eyes (see Personality Dolls chapter).

The Princess Elizabeth mold was also used for many more dolls, including McGuffey Ana, Kate Greenaway, Flora McFlimsey, and Snow White (see Character Dolls chapter). McGuffey Ana and Snow White were being marketed circa 1938-39, Flora joined the line circa 1940-41, and Kate was featured in the Alexander catalog in 1942.

The McGuffey Ana dolls came in a variety of sizes including 9" (Little Betty Mold with painted eyes), 11" (Betty closed mouth mold), 14", 16", 18", 20", and 24". The larger dolls were made with the Princess Elizabeth mold. All of the dolls had their hair styled in pigtails. The McGuffey Ana doll was a good seller and remained in the Alexander line of dolls off and on for many years.

Flora McFlimsey was sold in 15" and 21" sizes in the N. Shure Co. catalog of 1941-42. This doll is distinguished because of the freckles on her nose. She was made from the Princess Elizabeth mold.

The Kate Greenaway dolls had golden yellow hair styled in curls and their clothing was fancier than the dresses worn by McGuffey Ana. They came in sizes of 15", 16", 20", and 24". They also were made with the open mouth Princess Elizabeth mold.

Besides these dolls, the Princess Elizabeth mold was used for a variety of other dolls including flower girls. These dolls were made to accompany the Alexander brides and bridesmaids dolls.

The Wendy Ann doll was another of the basic doll molds used from the late 1930s until the end of the composition doll era in 1947. Named for Madame Alexander's granddaughter, this doll began life as a 13" composition doll with a jointed waist. Eventually she was made in sizes of 13", 15", 18", 21", and 22". The larger dolls did not include the bending waist. The basic Wendy doll also became Scarlett O'Hara, Fairy Princess, Madelaine du Bain, Alice in Wonderland, World War II service men and women, Miss America, various brides and bridesmaids, and Carmen. She was also the basic model used for the special 21" Portrait series circa 1946.

Wendy was featured as a 13" Alice in Wonderland doll in the 1939 Montgomery Ward Christmas catalog. The doll had sleep eyes, closed mouth, and a human hair wig. Her unusual jointed waist made her special. The larger dolls, produced later, did not have this feature. A Wendy Ann doll with molded hair was shown in the John Plain catalog that same year. She had painted eyes and a jointed waist.

In 1939, Madame Alexander added another personality doll to the market when the Sonja Henie doll was introduced. Some of these dolls used the Wendy Ann jointed waist body but most were made with a special Sonja Henie mold (see Personality Dolls chapter). The Sonja doll was still being marketed by Alexander in 1942 but she was then called the Skating Doll (evidently to avoid paying royalties to Sonja Henie).

In December 1939, the Madame Alexander Company began to profit from one of their most popular business decisions when the authorized Scarlet O'Hara doll was put on the market. This act coincided with the release of the David O. Selznick *Gone With The Wind* film, based on Margaret Mitchell's famous book. Through the years, the Scarlet dolls were made in a variety of styles and materials and remained a staple of the Alexander firm on and off for decades (see Personality Dolls chapter). These composition dolls were still being offered in the 1942 Alexander catalog but again, to avoid royalties, the dolls had been renamed "Southern Girl."

The Alexander company made beautiful baby dolls throughout the 1930s and 1940s. Most were called either Baby Genius or Baby McGuffey and came with composition heads, arms and legs, cloth bodies with criers, and sleep eyes. The Baby Genius dolls were made in styles that had molded hair and also were sometimes produced with wigs. An Alexander baby doll that appeared to be Baby Genius was featured in the Montgomery Ward 1939 Christmas catalog in a large 23-3/4" size priced at $4.69. She had molded hair and sleep eyes. In the 1942 Alexander catalog, the Baby Genius dolls came in sizes of 11", 14", 18", 24", and 29". All came with molded hair but the 18", 20", and 24" sizes could also be purchased with wigs. The same catalog offered Baby McGuffey dolls in 11", 14", 18", 20", and 24" sizes. All of these dolls came with wigs.

In the later 1940s, circa 1948-49, the Baby Genius dolls were made with hard plastic heads, latex arms and legs, and cloth bodies.

Besides the Baby Genius and Baby McGuffey dolls, other, more unique, baby dolls were sold from time to time. Included were the Pinkie, "Lov-Le-Skin," and Slumbermate dolls.

The Pinkie doll was advertised in the N. Shure Co. catalog for 1941-42. This doll had a new infant type face similar to the American Character "Little Love" doll then on the market. The head, arms, and legs were composition and the body was cloth. The doll had sleep eyes, closed mouth, and molded hair. It came in a 20" size in the N. Shure catalog but has been found in an 18" size.

The 1941-42 N. Shure catalog also pictured an Alexander baby doll called "Lov-Le-Skin." The author received one of these dolls for Christmas in 1942. The catalog's dolls came in 20" and 24" sizes. They had composition heads with sleep eyes and fully jointed bodies made of latex rubber. "Lov-Le-Skin" came with underwear (not a diaper), slip, organdy dress and matching bonnet, and booties. When the author retrieved her childhood doll from closet storage, she found the doll's body had deteriorated so much that she threw the doll away. This is probably what happened to most of these dolls, as they are nearly impossible to find.

The Slumbermate dolls with the eyes painted shut also date from the 1940s and were probably produced around the same time as similar dolls made by Ideal and Effanbee (circa 1945). The Alexander dolls had composition heads, hands and lower legs, and cloth bodies. The hair was molded and the features were painted (with closed eyes). The dolls were made in 12" and 21" sizes.

The Alexander Butch dolls could also be classified as baby dolls, although they were dressed in boy clothing. They were made with baby doll construction including composition heads, arms and legs, and cloth bodies. The 1942 Alexander catalog listed the dolls in sizes of 11", 14", and 18" sizes. All came with wigs. In the later 1940s, the 11" Butch doll was teamed with the 11" Baby McGuffey to be sold as a set of dolls called Bitsy and Butch. They were carried in the 1947 Montgomery Ward Christmas catalog and were priced at $9.69 for the pair.

Although the Madame Alexander firm continued to use Wendy Ann dolls as the basis for many other dolls, some of these dolls were extra special. Included were the dolls called Madelaine du Bain, Fairy Princess, Carmen, and the World War II service men and women (see Military chapter).

The Madelaine du Bain dolls had lovely clothing and many wore hooped skirts and pantaloons. Their human hair wigs were sometimes styled in curls. They came in sizes of 14", 18", and 20" and are circa late 1930s-1940.

The Fairy Princess dolls are also favorites of collectors. These dolls were featured in the 1942 Alexander catalog in sizes of 11", 15", 18", and 22". All had long, shoulder length blonde wigs and wore full length gowns accented with long necklaces.

The Carmen dolls were marketed by Alexander circa 1942, when movie star Carmen Miranda was so popular. Although these dolls were not called Carmen Miranda, there is no doubt that there was a tie-in to the star, who was originally from Brazil (see Personality Dolls chapter). The 1942 Alexander catalog offered the Carmen dolls in sizes of 7", 8", 11", 15", 18" and 22". The smaller dolls were made by using the Betty dolls while the larger dolls were based on Wendy Ann molds.

Also pictured in the 1942 Alexander catalog was an unusual set of two dolls called "Mother and Me." Included were a 15" mother doll and a 9" daughter doll, complete with sets of matching cloth-

ing. The larger doll was made from a Wendy mold with sleep eyes, closed mouth, and wig. The smaller doll was a Little Betty model with painted eyes, closed mouth, and a mohair wig. Several different sets of these dolls were marketed by Alexander. One included thirty-four pieces, counting the dolls. These dolls were probably also on the market in 1941.

7-1/2" Alexander composition Tiny Betty Fairy Princess. She is all original, in mint condition, and has her original box ($450-500 in this condition, with box, and in this outfit). *Doll and photograph from the collection of Veronica Jochens.*

Although most of Alexander's line of composition dolls used the same basic models with changes made to their clothing and wigs to make them take on the look of various characters, some special dolls from the era were made to be used for only one doll. Included were the Sonja Henie and Jane Withers dolls as well as the Jeannie Walker doll. Jeannie Walker was pictured in the 1941-42 N. Shure Co. catalog in the 18" size; she also came in a 14" size. These dolls had a patented feature that enabled them to walk (with help). Jeannie dolls had sleep eyes, closed mouth, and a human hair wig. They were not sold for as many years as most of the other Alexander dolls, so they are special to collectors when found with their original clothing.

The Margaret-faced dolls were the last great composition dolls marketed by the Alexander firm. Following their earlier practice, the company produced a variety of dolls with different names, all using the same basic doll. The most collectible of these dolls is the Margaret O'Brien doll from 1946-47 (see Personality Dolls chapter). Other important dolls using the Margaret face include Margaret Rose (see Personality Dolls chapter), Karen Ballerina, Alice in Wonderland, and Enchanting Fairy Queen. These dolls were pictured in the Montgomery Ward and Sears Christmas catalogs for 1947. The Karen Ballerina was listed in sizes of 15", 18", and 21", while Alice in Wonderland came in 14-1/2", 18", and 21" sizes. The Fairy Queen came in only two sizes of 14-1/2" and 17-3/4". She not only wore a crown on her head, she also had wings and carried a magic wand. The Margaret mold continued to be

Alexander Doll Co.

14" composition Alexander Princess Elizabeth with wrist tag, circa 1941-42. She has sleep eyes, open mouth with teeth, a mohair wig, and is fully jointed. Her coat is trimmed in fur and is made of a lightweight wool. She has a matching hat and muff. Her wrist tag reads "Created/ By/ Madame/Alexander/New York." The 1942 Alexander catalog listed Jeannie Walker and Kate Greenaway dolls dressed in fur trimmed coats and hats similar to this one and the shape of the tag matches those of the Jeannie Walker dolls circa 1941-42 (not enough examples to determine a price). *Doll and photograph from the collection of Veronica Jochens.*

Above left:
The Sears Christmas catalog for 1937 pictured what appears to be the doll collectors call "Special Girl." This doll came in 16" and 24" sizes and was priced at $2.98 and $4.98 each. She had sleep eyes, closed mouth, a real hair wig, and cry voice. She was called "Pig Tails" in the catalog. *From the collection of Marge Meisinger.*

Above center:
16" Alexander doll called "Special Girl" by collectors and pictured in the Sears catalog for 1937. She has a composition head, shoulder plate, full arms and nearly full legs, with a cloth body and upper legs. The doll has sleep eyes, closed mouth, and a human hair wig styled in pig tails. She also has a crier as described in the catalog. The back of her head is marked "ALEXANDER." She is all original and is wearing the same dress as pictured in the catalog ($350-400+).

13" composition Alexander rare "Princess" doll using Betty face. She has sleep eyes, closed mouth, human hair wig, and is fully jointed. She is all original wearing a taffeta dress that was probably blue when new, but is now lavender. Her bonnet is black velvet. The dress is tagged only "Princess" ($800+). *Doll and photograph from the collection of Veronica Jochens.*

22" composition Alexander Princess Elizabeth. She has sleep eyes, open mouth with teeth, mohair wig, and is fully jointed. She is all original wearing a white polka dot dress, red coat, hat trimmed with fur and feather, and matching muff. Even though she was made with the Princess Elizabeth head, it is not known who she was supposed to represent as she is tagged only "Madame Alexander" ($850+). *Doll and photograph from the collection of Veronica Jochens.*

14" composition Alexander McGuffey Ana wearing a very rare outfit. This doll still has her box and tag and is in mint condition. Her dress is made of plaid taffeta and cotton organdy. Her "School House" tag reads "I Am/MCGUFFEY-ANA/and MADAME ALEXANDER/has made me look like my mommie/did when she went to school./I hope I will make you/very happy" (not enough examples to determine a price). *Doll and photograph from the collection of Veronica Jochens.*

McGUFFEY ANA

Little McGuffey Ana, adorable and sweet as can be. She has real flaxen hair, with ribbon tied braids and cute curls. Dressed in bright Red Scotch plaid, with apron of White organdy, lace trimmed; 15-in. doll has no hat.

The John Plain catalog pictured Alexander's McGuffey Ana doll in their catalog for 1940. The composition dolls were available in sizes of 15", 20", and 24". The 15" doll came with no hat. *From the collection of Marge Meisinger.*

16" composition Alexander McGuffey Ana, circa 1940-41. She has sleep eyes, open mouth with teeth, human hair wig in braids, and is fully jointed. She is all original, complete with tag. Her dress tag reads "McGuffey Ana/Madame Alexander, N.Y. U.S.A." Her wrist tag says "An/Alexander/Product" on one side and "Created By MADAME ALEXANDER/NEW YORK" on the other. The back of her neck is marked "Princess Elizabeth/Mme. Alexander." Her hat and pinafore are like those on a McGuffey doll from 1941 and her tag is the same as those on the Jeannie Walker dolls circa 1941-42 ($600).

Left:
The N. Shure Co. pictured an Alexander Flora McFlimsey doll for sale in their 1941-42 catalog. She could be purchased in 15" or 21" sizes. *From the collection of Marge Meisinger.*

Right:
14" composition Alexander Flora McFlimsey similar to the one pictured in the N. Shure Co. catalog for 1941-42. The Princess Elizabeth mold was used for this doll. She has sleep eyes, open mouth with teeth, painted freckles, human hair wig, and is fully jointed. The doll is all original and is wearing a dress tagged "Flora McFlimsey of Madison Square, Madame Alexander N.Y. USA All Rights Reserved" ($700-800+). *Doll from the collection of Lois Jakubowski. Photograph by Robert Jakubowski.*

11" composition Alexander McGuffey Ana in a rare outfit (color and hat) and with the Betty mold. She has sleep eyes, closed mouth, human hair wig in braids, and is fully jointed. Her dress is tagged "McGuffey Ana" and she is all original. The necklace has been added ($350+). *Doll and photograph from the collection of Veronica Jochens.*

The Alexander McGuffey Ana doll was offered in a "Trousseau" set in the 1941 Carson Pirie Scott & Co. catalog. The doll was 13" tall and came with an organdy dress, shoes, stockings, pinafore apron, straw hat, handkerchief, spun rayon coat, pocketbook, and batiste nightgown with lace trim. Separate dolls could be purchased in sizes of 13", 16", 20", and 24". *From the collection of Marge Meisinger.*

15" composition Alexander Kate Greenaway, circa 1942. She has sleep eyes, open mouth with teeth, mohair wig, and is fully jointed. The doll was made with the Princess Elizabeth mold. The dresses on the Greenaway dolls were tagged "Kate Greenaway Madame Alexander New York All Rights Reserved" ($650-750). *Doll from the collection of Marge Meisinger. Photograph by Carol Stover.*

20" composition Alexander unidentified doll. She has the gold mohair wig as do the other Kate Greenaway dolls but her clothing is not tagged "Kate Greenaway." Instead, it is tagged "Madame Alexander." The doll has sleep eyes, open mouth with teeth, and is fully jointed. She is all original and wears the same style shoes as the Greenaway dolls so she is from the same period. Her wrist tag reads "Created by/ Madame Alexander New York NY" (not enough examples to determine a price). *Doll and photograph from the collection of Veronica Jochens.*

15" composition Alexander Kate Greenaway wearing rare colored dress. She is all original and in near mint condition. She has a gold color mohair wig (all of the Greenaway dolls have this color hair). She wears a "Kate Greenaway" tagged dress ($850 in this condition). *Doll and photograph from the collection of Veronica Jochens.*

Left:
24" composition Alexander Kate Greenaway doll in a rare size. She is all original in her tagged dress. These dolls were listed in the Alexander catalog for 1942 in sizes of 13", 15", 16", 20", and 24" ($1,000+). *Doll and photograph from the collection of Veronica Jochens.*

Right:
18" composition Alexander Flower Girl, circa 1940s. This doll also uses the Princess Elizabeth face. She has sleep eyes, open mouth with teeth, mohair wig, and is fully jointed. She is all original with her wrist tag. Her nylon dress is tagged "Madame Alexander" (not enough examples to determine a price). *Doll and photograph from the collection of Veronica Jochens.*

The Montgomery Ward Christmas catalog for 1939 featured an Alexander Wendy Ann "Alice-in-Wonderland" doll. The copy stated "A special joint at the waist makes her more flexible than other dolls." Madame Alexander was listed as the designer of the clothing. Alice was 13" tall. *From the collection of Marge Meisinger.*

13" composition Alexander "Alice-in-Wonderland" as pictured in the 1939 Montgomery Ward Christmas catalog. She has sleep eyes, closed mouth, human hair wig, and is fully jointed including a joint at the waist. She is marked on the back "Wendy-Ann/ Mme Alexander/ New York." Her dress tag reads "Alice in Wonderland by Madame Alexander, New York, All Rights Reserved." She is all original except for her replaced shoes. This doll was given to the author for Christmas in 1939 ($200-250 with some wear and replaced shoes).

The composition Alexander Wendy Ann with molded hair was advertised in the John Plain 1939 catalog. She was 13-1/2" tall. *From the collection of Marge Meisinger.*

13-1/2" composition Alexander Wendy Ann, circa 1939. She has painted features, molded hair, and is jointed at the shoulders, hips, and waist. She is marked on the back "Wendy-Ann/ Mme Alexander/New York." The doll has been redressed in a plaid taffeta dress and hat. Her shoes and socks have also been replaced. The molded hair Wendy is much harder to find than the wigged doll ($250-300 in this condition).

Alexander Doll Co. 17

14" composition Alexander Wendy Ann, circa 1940-41. She has sleep eyes, closed mouth, human hair wig, and is in mint condition. She is all original wearing a white dotted swiss dress with red ribbon trim. Her felt hat is trimmed with flowers and fruit. The doll also has her wrist tag and original box. The tag reads "Created/by/Madame/Alexander/New York" (not enough examples to determine a price for MIB doll). *Doll and photograph from the collection of Veronica Jochens.*

13" composition Alexander Wendy Ann, circa late 1930s. She has sleep eyes, closed mouth, human hair wig, and is fully jointed (including the waist joint). She carries the "Wendy-Ann" markings on her back and her dress is tagged "Wendy-Ann/MME ALEXANDER N.Y/ALL RIGHTS RESERVED." She is all original except that she is missing her black lace gloves ($350 not mint).

Left:
The Montgomery Ward Christmas catalog pictured a 23" Alexander baby doll in 1939. Although the doll isn't named, it is likely a Baby Genius doll. The head and 3/4 of the arms and legs were composition while the body and the rest of the limbs were cloth. The doll had sleep eyes, molded hair, and a cry voice. It sold for $4.69. *From the collection of Marge Meisinger.*

Right:
11" composition Alexander Baby Genius, circa 1940s. The doll has sleep eyes, closed mouth, molded hair, composition head, lower arms and legs, and a cloth body and upper limbs. She is wearing her original white dress and matching bonnet, both trimmed with pink ribbon. The doll is missing her panties and has replaced shoes and socks. She is marked on her head "MME ALEXANDER" ($100 in this condition).

14" Alexander Baby McGuffey, circa 1940s. She has sleep eyes, closed mouth, mohair wig, composition head and lower limbs, and cloth body and upper limbs. The doll comes with her original tag and box. Her tag reads "Created By/Madame Alexander" (MIB $500+). *Doll and photograph from the collection of Veronica Jochens.*

The popular Alexander "Pinkie" doll was featured in the 1941-42 N. Shure Co. catalog. The 20" doll was dressed in a christening dress and bonnet. *From the collection of Marge Meisinger.*

20" Alexander "Pinkie" doll, as pictured in the N. Shure Co. Catalog for 1941-42. The doll has sleep eyes, closed mouth, molded hair, composition head, composition lower arms and legs, and a cloth body and upper limbs. The doll came with a cry box and had the look of a very young infant. Pinkie appears to be wearing the same christening dress and bonnet pictured in the catalog. This doll is all original and comes with her original box and tag (MIB $800). *Doll and photograph from the collection of Veronica Jochens.*

20" Alexander Baby McGuffey, circa early 1940s. She has sleep eyes, closed mouth, mohair wig, composition head, composition hands and lower legs. Her body, upper arms, and legs are cloth. A cry box is housed in her back. The doll wears her original faded blue dress and pink pinafore and bonnet. Her head is marked "Mme Alexander." Her dress tag reads "Baby McGuffey/Madame Alexander/N.Y.U.S.A/All Rights Reserved" ($300-350).

Alexander Doll Co. 19

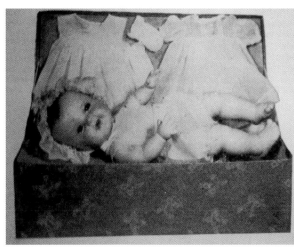

The 1941-42 N. Shure Co. catalog also pictured an Alexander baby doll called "Lov-Le-Skin" Baby, listed in sizes of 20" and 24". The doll had a composition head and a jointed body made of latex. She had molded hair and sleep eyes. The author received the 20" size for Christmas in 1942. The early latex "skin" did not wear well and most of these dolls were discarded after a few years. They are now very hard to find. *From the collection of Marge Meisinger.*

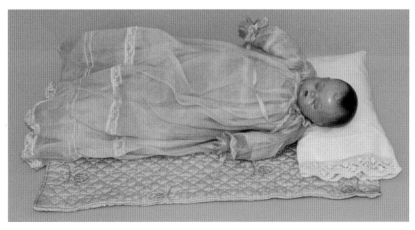

12" Alexander Slumbermate baby, circa 1943-45. She has eyes painted shut, closed mouth, and molded hair. Her head, lower legs and hands are composition while her body and upper limbs are cloth. She is marked "Alexander" on the back of her head. Her long lace-trimmed dress and slip are original ($225+).

15" composition Alexander Madeline, circa 1940. She has sleep eyes, closed mouth, human hair wig, and is fully jointed. She was made with the Wendy Ann mold. She wears a printed cotton dress with hoop skirt ($500+). *Doll from the collection of Marge Meisinger. Photograph by Carol Stover.*

18" composition Alexander Madeline du Bain, circa 1940. She was also made using the Wendy mold. The doll is rare in this size and in this costume (not enough examples to determine a price). *Doll and photograph from the collection of Veronica Jochens.*

20" composition Alexander Madeline du Bain, circa 1940. She was made using the Wendy mold and is all original with her original tag. She wears a pink organdy dress, black straw hat trimmed with flowers, hoop skirt, and pantaloons, and is in mint condition (Mint $1,000+). *Doll and photograph from the collection of Veronica Jochens.*

18" composition Alexander Fairy Princess, circa early 1940s. She was made from the Wendy mold and has sleep eyes, closed mouth, human hair wig, and is fully jointed. She wears a satin dress with delicate embroidered panels down the front, gold leatherette shoes, gold sequin tiara, and long gold necklace. The Alexander catalog for 1942-43 listed the Fairy Princess doll in sizes of 11", 15", 18", and 22" ($1,000 in excellent condition). *Doll and photograph from the collection of Veronica Jochens.*

11" composition Alexander Fairy Princess, circa 1942-43. She wears a pale blue satin gown with organdy slip and sequin tiara ($550-650). *Doll from the collection of Marge Meisinger. Photograph by Carol Stover.*

14" and 9-1/2" Mother And Me Alexander composition dolls, circa early 1940s. The dolls were made using the Wendy and Betty molds. Both dolls have closed mouths and mohair wigs, while the larger doll has sleep eyes and the small doll's eyes are painted. Both dolls are marked "Mme Alexander/New York." Some of the clothes made for the Mother And Me sets were labeled "Mother and Me," others had "Mme. Alexander" tags, and some apparently did not have labels at all. *Dolls and photograph from the collection of Veronica Jochens.*

Alexander Doll Co. 21

Below:
15" composition Alexander "Jeannie Walker," circa 1941-43. She is MIB with her wrist tag. She has sleep eyes, closed mouth, mohair wig, and a jointed walker body. She wears a red blouse, navy pleated skirt, and matching hat. She is marked "Alexander" on torso (sold at auction for $1,732). *Photograph courtesy of Frashers' Doll Auctions, Inc.*

Here Comes "Jeannie Walker"
Madame Alexander's Cleverest Creation
She Actually Walks!
This New and Patented Feature makes her the *only* Doll with a really natural walk

18-inches tall. Fine quality all-composition body, fully jointed, with glass sleeping eyes, with eyelashes. Curled human hair wig. Has a patented feature that enables it to actually walk. Placing the hands on the doll as illustrated, and applying a slight pushing, will cause this sensational new creation to walk just as freely and easily as any child. Dressed in finest quality assorted dresses with complete undergarments, shoes and stockings and cute straw hat.
No. 36N294. Each.................................. 4.35

The N. Shure Co. catalog for 1941-42 pictured the Alexander "Jeannie Walker" doll in the 18" size. The copy stated that the doll had a patented feature that enabled the doll to actually walk (with help).
The Alexander catalog from 1942-43 listed the dolls in 14" and 18" sizes. *From the collection of Marge Meisinger.*

20" composition Alexander Jeannie Walker. She wears her original dress with pleated skirt and matching hat and is in excellent condition ($900+). *Doll and photograph from the collection of Veronica Jochens.*

The "Mother And Me" packaged set came with an assortment of matching clothes for both dolls. Pictured are cotton dresses, play clothes, apron dresses, and gowns. The 1942-43 Alexander catalog pictured the set in a package similar to those sold with the earlier Sonja Henie dolls. The catalog stated that there were thirty-four pieces in the set including the dolls (not enough examples to determine a price). *Dolls and photograph from the collection of Veronica Jochens.*

18" composition Alexander doll made using the Wendy mold. She may be from the Southern Girl series, circa 1940s. The doll has sleep eyes, closed mouth, mohair wig, and is fully jointed. Her dress came tagged as only "Madame Alexander/New York." She wears a gown of blue and white cotton trimmed with white eyelet and red ribbon, and a blue straw hat trimmed in flowers. Her underwear includes pantaloons, and a hoop slip ($1,000+). *Doll and photograph from the collection of Veronica Jochens.*

The 1944 Sears Christmas catalog pictured a beautifully dressed Alexander Bridesmaid doll. She came in sizes of 15", 18", and 22-1/2", priced at $6.59, $9.98. and $11.98 each. *From the collection of Marge Meisinger.*

Right:
22" Alexander composition Bridesmaid, as shown in the Sears 1944 catalog. The doll uses the Wendy mold. She has sleep eyes, closed mouth, mohair wig, and is fully jointed. The doll is all original complete with the flowers in her hair and her bouquet. Her gown is made of a yellow sheer rayon marquisette trimmed in ribbon (Mint $750+). *Doll and photograph from the collection of Nancy Roeder.*

21" composition Alexander doll made with the Wendy mold. She may also be from the Southern Girl series, circa 1940s. She wears a blue and white striped dress trimmed with rickrack, flowers, and ribbon, and a flower trimmed straw hat. She has a hoop skirt slip and pantaloons ($1,000+). *Doll and photograph from the collection of Veronica Jochens.*

Alexander Doll Co. **23**

Madame Alexander composition Bridal Party dolls were featured in the Sears 1945 Christmas catalog. The bridesmaid doll came in sizes of 15", 18", and 21-1/2", priced at $6.84, $9.98, and $12.45. The bride was listed in 17" and 22" sizes at $10.39 and $12.79 each. Both dolls were made using the Wendy face. The flower girl probably used the Princess Elizabeth face and came in sizes of 16", 20", and 24", priced at $8.98, $12.45, and $14.98 each.

11" Alexander Bridesmaid similar to the dolls advertised in the 1944 Sears catalog. She has sleep eyes, closed mouth, mohair wig, and is fully jointed. The doll has a fuller face than the Wendy dolls and is probably made from the Betty mold. She is all original, wearing a rayon marquisette dress trimmed with ribbon and flowers. She carries a flower trimmed purse and still has the original flowers in her hair ($650 in this design and dress). *Doll and photograph from the collection of Veronica Jochens.*

Right:
The Sears Christmas catalog for 1946 also pictured a page of Alexander composition dolls. Bride and bridesmaid dolls were offered in new costumes. The bridesmaids came in sizes of 15", 18", and 21-1/2" and were priced at $9.98, $12.98, and $15.98 each. The bride was listed in sizes of 17" and 21-1/2" and cost $15.98 and $17.98 each. Both dolls were made using the Wendy faces. The third doll was the composition Margaret O'Brien doll. She came in sizes of 14-1/2", 18", and 21", priced at $7.79, $10.98, and $13.98. *From the collection of Betty Nichols.*

17" composition Alexander Margaret O'Brien, circa 1946-47. She has sleep eyes, closed mouth, mohair wig, and is fully jointed. She is all original ($1,100-1,300). *Doll from the collection of Lois Jakubowski. Photograph by Robert Jakubowski.*

The Sears Christmas catalog for 1947 also featured a page of Alexander dolls. Included were Alice in Wonderland, Enchanting Fairy Queen, Margaret O'Brien, Bitsey and Butch, and a baby doll with latex skin. The Fairy Queen came in sizes of 14-1/2" and 17-3/4", priced at $9.98 and $14.98. Margaret was listed in sizes of 14-1/2", 18", and 21", priced at $7.49, $9.98, and $12.98 each.

The Montgomery Ward Christmas catalog for 1947 featured four composition Alexander dolls. Included were Margaret Rose (15", 18", 21" sizes), Hulda (18" size only), Karen Ballerina, and Sleeping Beauty (15", 18" and 21" sizes), The Karen Ballerina came in sizes of 15", 18", and 21" and was priced at $8.99, $10.79, and $13.79.

18" composition Alexander Karen Ballerina, circa 1947. She has sleep eyes, closed mouth, braided wig, and is fully jointed. The doll is marked "Alexander" on her head and body. She is original except for flowers missing at her waist ($600-700 with some wear to costume).

Alexander Doll Co. 25

18" composition Alexander Alice in Wonderland, as pictured in the Sears Christmas catalog in 1947. She has sleep eyes, closed mouth, mohair wig, and is fully jointed. She is marked "Alexander" on her back. Her dress tag reads "Alice In Wonderland/ Madame Alexander N.Y.U.S.A./TM 304,488." She has the Margaret face and is all original. The Sears catalog listed the Alice dolls in sizes of 14-1/2", 18", and 21", priced at $8.98, $10.98, and $13.98 ($600).

21" Alexander baby doll (probably Baby Genius), circa late 1940s. She has a hard plastic head, sleep eyes, closed mouth, molded hair, and arms and legs of stuffed latex rubber. Her body is cloth. The back of her head is marked "Alexander." She is all original except for her shoes and socks. This type body was used on the large baby doll pictured in the Sears 1947 Christmas catalog. That doll had a wig and came in sizes of 18" and 20-1/2". The "skin" darkened with age ($75+ in this condition).

11" Alexander Bitsey and Butch, circa 1947. The dolls have sleep eyes, closed mouths, mohair wigs, composition heads, hands, and legs, and cloth bodies and upper limbs. They are marked on their heads "Mme Alexander." Butch's clothing is marked "Butch"/Madame Alexander NYUSA/All rights Reserved." Bitsey's tag is the same except for "Bitsey." Both dolls are original. The copy in the Sears 1947 Christmas catalog stated that the dolls were "New." The pair of dolls was priced at $9.69 in the catalog ($150-175 each).

American Character Doll Company

The American Character Doll Company of New York City had its beginning in 1919. The firm produced many popular composition dolls during the 1920s, 1930s, and 1940s, manufacturing both baby and little girl composition dolls. Several of the early dolls were marketed under the trade name "Petite."

Many of the firm's baby dolls from the mid-1920s to the late 1930s carried the "Bottletot" name. Several different designs of these dolls were produced. The first Bottletot dolls had one hand molded to hold a bottle; the later dolls did not retain that feature. The earlier dolls were manufactured with composition heads, shoulder plates, and arms, and cloth bodies and legs. These dolls were made in 13" and 16" sizes. The later dolls were "drink and wet" models and had bodies made of rubber. In 1938, these dolls were issued in 11" and 15" sizes. Sears sold the dolls in suitcases that included wardrobes. Bottletot dolls were still being marketed by American Character in 1941. This later baby doll was also a "drink and wet" model and came in sizes of 11", 13" and 15".

Earlier, circa 1934, American Character had begun production of a wetting doll called "Wee Wee." The firm was taken to court by the Effanbee Co. (Fleischaker and Baum) because they thought the new doll's name infringed on the copyright for their Dy-Dee "drink and wet" doll. After American Character lost the court case, their wetting dolls were renamed Bottletot.

By 1936, the company's "drink and wet" dolls carried the trade name "Marvel-Tot." These dolls had an extra feature which enabled them to smile or pout with a turn of their heads. They were advertised by Sears in 1937 in 10-1/2" and 12-1/2" sizes. The babies were priced from $2.98 to $4.98, depending on the size of the doll and the layette. All came with molded hair and sleep eyes.

Toddle Tot and Happy Tot were other trade names used by American Character for baby dolls made in the late 1920s and early 1930s. These dolls were also made in several different designs and sizes. Both dolls had cloth bodies but the Toddle Tot dolls, advertised in 1930, were similar to other composition Mama dolls of the period. These dolls came in sizes of 18", 21", and 24" and were priced at $3.48, $4.69, and $5.95 each.

American Character also produced a baby doll with a rubber body that was not a "drink and wet" doll. The company called this new rubber product Flex-o-flesh. The doll, circa 1933, was called Toodles and came in sizes of 15", 16-1/2", and 18-1/2". Smaller sizes of 10-1/2" and 12-1/2" were also made and named Little Toodles. The dolls were advertised in a Montgomery Ward catalog in 1933 and had composition heads, molded hair and sleep eyes. The doll's eyes stayed open when she was put to bed unless her head was turned to the side. A later Baby Toodles (1938) was made with a Paratex head, arms, and legs (see next paragraph), and a cotton stuffed body. This doll *was* a "drink and wet" style doll, which was unusual since she had a soft body. She was made in 15" and 18" sizes and appeared to be the same doll as Baby Sunshine.

In the mid-1930s, the American Character firm began experimenting with the material they called Paratex. This was a hard rubber substance and several of their dolls in this period were advertised as being made of Paratex. Included were Baby Sunshine and Sally Jane dolls. Sally Jane came in 15" and 19" sizes in 1937. Paratex held up much better than composition and it is too bad more companies did not use this product to make their dolls.

Other baby dolls manufactured by American Character included several models of Chuckles, made in the "Mama" doll style, in 1943. Another more unusual baby doll, marketed in the same year, was the Little Love doll, which was modeled to look like a very young infant.

Some of the most popular little girl dolls marketed by American Character were the Sally dolls. These all composition dolls were made to compete with the Effanbee Patsy. Introduced in 1930, Sally was approximately 12" tall, had painted features, and molded hair. This same model was also used for a boy doll when marketed as a boy and girl twin set of dolls. The firm also produced a Sally Joy doll with a composition shoulder plate, composition head and limbs, and a cloth body. She had sleep eyes and a wig. This type doll was advertised in the Sears 1930 catalog with molded hair in a 19" size for $3.48.

By the late 1930s and early 1940s, American Character was producing all composition little girl dolls dressed in a variety of outfits. Although the dolls were modeled as little girls, their costumes included W.A.V.E. uniforms, bridal gowns, skating ensembles, and formals as well as the more traditional little girl dresses. These dolls featured sleep eyes and wigs.

Although all American Character products are very much in demand, their most collectible dolls are probably the character designs. These include the Puggy doll, based on the character created by Charles Twelvetrees. Although this all composition chubby boy doll was introduced in the late 1920s, he was still being marketed in the 1930s. Puggy was 12" tall with painted features and molded hair. He was sometimes teamed with the Campbell Kid little girl and sold as part of a set. The composition Campbell Kid dolls were marketed in American Character's line in 1929-30. They were 12-1/2" tall and were based on the characters in the Campbell Soup advertisements (see Character Dolls chapter).

American Character was also responsible for one composition celebrity doll in 1935. The doll represented Carol Ann Beery, the adopted daughter of famous film star, Wallace Beery (see Personality Dolls chapter).

Like Arranbee, the American Character Doll Company usually played "follow the leader" in developing its line of dolls each year. The exception was when they issued their line of dolls made of Paratex. Present day doll collectors can only wish that this doll material had been more successful so that more of these perfectly preserved dolls would now be available for purchase.

American Character Doll Company 27

Spiegel advertised a Bottle Tot doll in their 1941 catalog. These "drink and wet" rubber dolls came in 11", 13", and 15" sizes and sold for $1.98-$3.98 each. *From the collection of Marge Meisinger.*

16" Bottletot made by American Character, circa 1930. The doll has a composition head, shoulder plate, full arms, and lower curved legs with a cloth body. She has sleep eyes, painted hair, and an open mouth. The right hand is molded to hold a bottle. Marked "Petite/Amer. Char. Doll Co." on shoulder plate. Replaced bottle and shoes, otherwise original ($125+).

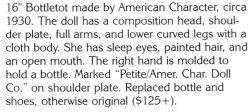

Catalog page from the 1937 Sears catalog featuring an American Character Bottletot doll. This was a "drink and wet" model with a rubber body and head. The dolls had sleep eyes and molded hair. They came in sizes of 10-1/2" and 14-1/2" and sold for $1.89 or $3.98 complete with layettes. *From the collection of Marge Meisinger.*

18" tall Toodles made by American Character, circa early 1930s. The doll's head is composition with molded hair, sleep eyes, and open mouth. The jointed body is rubber. The doll is identified as an American Character product on the back of its neck. This doll has been re-dressed ($50 in this condition)

American Character Doll Company

The American Character Toodles dolls were featured in the Montgomery Ward catalog for 1933. The doll's body, arms, and legs were made of "Flex-o-flesh" rubber. They came in sizes of 15", 16-1/2", and 18-1/2". Smaller versions were offered under the Little Toodles name in 10-1/2" and 12-1/2" sizes. *From the collection of Marge Meisinger.*

Sears featured the American Character Marvel-Tot "drink and wet" doll in their catalog for 1937. The doll could also smile and pout with a turn of her head. She came in 10-1/2" or 12-1/2" sizes. *From the collection of Marge Meisinger.*

16" Baby Sunshine made by American Character, circa 1938. She has sleep eyes with lashes, open mouth with two teeth, and molded hair. Her head is jointed to a neck piece. The head, lower arms, and legs are made of Paratex while the body and upper arms and legs are cotton stuffed. The covering over these parts is a water resistant oilcloth type material. The doll is a drink and wet model and a hole in her bottom allows the water to drain out. The Paratex material has stood the test of time and there is not a crack on her. She is all original. The doll is not marked but she came with her original box ($250+ original with box).

Left:
The lid of the box for Baby Sunshine reads "It toddles, drinks, wets, cries, sleeps and is washable. It is made of the new unbreakable washable Paratex, (a new hard rubber material.) … long-lasting, Petite dolls."

American Character Doll Company 29

The Sears catalog in 1938 featured a baby doll called Baby Toodles that is apparently the same doll as Baby Sunshine. The doll came in 15" and 18" sizes and was priced at $2.98 to $4.98. This doll was a "drink and wet" doll with a cotton stuffed body. Although the trade name Paratex is not mentioned, the doll had a hard rubber head and arms and legs. *From the collection of Marge Meisinger.*

Other American Character dolls made of Paratex were advertised by Sears in their 1936 catalog. The little girl Sally Jane dolls came in two styles. One was completely made of Paratex while the other model had a soft rubber head which could smile, pout, or pucker. The dolls came in sizes of 15", 17" and 19". *From the collection of Marge Meisinger.*

12-1/2" all composition Sally made by American Character, circa early 1930s. She has molded hair, painted features, and is fully jointed. Marked on her back is "Sally/A Petite/Doll." She is wearing her original dress, one piece underwear, and shoes and socks. She is missing her hat. This doll was produced by American Character to compete with Effanbee's Patsy dolls ($175-200).

19" all original Chuckles made by American Character, circa 1943. She has a mohair wig, sleep eyes with lashes, and an open mouth with two teeth. Her head, lower arms, and lower legs are composition while her body, upper arms, and legs are cloth. She is marked "Am.Char. Doll" on the back of her head. She wears her original dress, bonnet, one piece underwear, and shoes and socks ($125).

American Character Doll Company

18" Little Love dolls made by American Character, circa 1942-43. The dolls were supposed to represent two-month-old babies. They have composition heads and hands with cloth bodies and legs, molded hair, sleep eyes, and closed mouths. The use of less composition and more cloth in the manufacture of dolls was a reflection of the shortage of materials during World War II. Marked on the back of the flange heads is "AM. CHAR. DOLL." The tag reads "Little Love/A Petite Baby." These dolls are all original. They were received as Christmas gifts by two sisters, Bonnie McCullough and Marilyn Pittman, circa 1942 (tagged with bonnet $275, other $225). *From the collections of Bonnie McCullough and Marilyn Pittman. Photograph by Marilyn Pittman.*

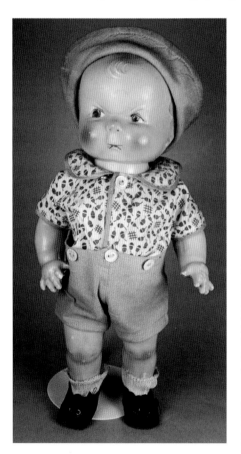

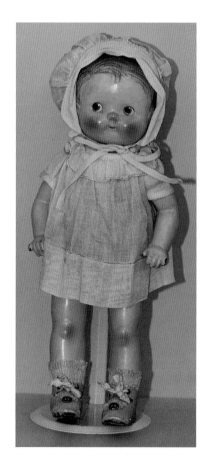

Left:
12" all composition Puggy made by American Character, circa early 1930s. The all jointed doll has molded painted hair, painted eyes, and a closed mouth. He was based on the characters drawn by Charles Twelvetrees. The doll is marked "A/Petite/Doll" and is all original ($500). *Doll from the collection of Marge Meisinger. Photograph by Carol Stover.*

Right:
12-1/2" composition Campbell Kid made by American Character, circa 1930. Molded hair (repainted in this example), painted eyes, and closed mouth. Marked on the back is "A Petite Doll." The clothes appear to be original. The dolls were based on the Campbell Kid characters used to advertise Campbell Soup ($275+ for this example) *Doll courtesy of Linda Friend of Friendly Antiques.*

Arranbee (R & B)

The Arranbee Doll Company (R & B) was in business from the 1920s until 1958. This New York firm is best known for its beautiful composition and hard plastic dolls of the 1930s, 1940s, and 1950s. The company marketed a wide range of baby dolls during their history, including the popular Dream Baby, which featured a bisque head during the 1920s. The firm continued to use the "Dream Baby" trade name for many of its composition baby dolls of the 1930s.

One of the company's most innovative baby dolls was the Drink 'n Babe doll, produced in 1935. The all composition jointed baby doll came with a bottle that could be filled with a special "milk" used to feed the doll. After feeding, the liquid reappeared in the bottle when it was held upright.

Another interesting baby doll from the period was the Ink-u-Bator Baby from 1936. This 7" all composition baby doll came in an incubator, complete with accessories. The idea of a baby doll in an incubator probably developed because of all the publicity given to the Canadian Dionne Quintuplets. These very tiny girls were placed in incubators shortly after they were born, which probably saved their lives. The same model doll was also used in an unauthorized set of Quintuplets to cash in on the popularity of the Dionne Quints.

The other Arranbee babies followed the trend of the day and were designed with composition heads, arms and legs, and cloth bodies. These dolls usually featured sleep eyes and came with either molded hair or wigs. Most of the dolls had criers in their stomachs. Some of the trade names included Little Angel and Dream Baby.

The Arranbee little girl dolls from the 1930s were very attractive as well, but most of the designs were similar to dolls being marketed by other companies. The nicest all composition dolls had sleep eyes and wigs and were dressed in quality clothing. Many of these dolls were called Nancy or Nancy Lee, but in one line, attributed to Arranbee, they were named Rosalie. Sears advertised these dolls in their 1938 and 1939 catalogs. Rosalie was described as a teen-ager and was probably produced to compete with Ideal's Deanna Durbin dolls. The cloth bodied Debu'Teen was a similar doll marketed by Arranbee during the same period. Although most collectors refer to Debu'Teen dolls as those with composition heads, composition arms and legs, and cloth bodies, other R & B dolls have been found bearing the Debu'Teen tags as well. These include 11" all composition dolls with wigs and painted eyes and 14" all composition dolls with wigs and sleep eyes. It may be that, like the Nancy and Nancy Lee trademarks, this name was used on a wide variety of dolls.

Another unique line of girl dolls, circa late 1930s, is known as the Southern Series. These dolls were dressed in long gowns and have an old-fashioned look. They were made in 14" and 17" sizes and may have been produced to compete with the Alexander *Gone With the Wind* Scarlet O'Hara dolls.

The Arranbee firm also marketed a line of chubby Nancy dolls that resemble the Alexander Princess Elizabeth dolls. These all composition Nancy dolls have sleep eyes, wigs, and extra nice clothing. They were marketed circa 1940.

Besides these special little girl composition dolls, the company marketed many less expensive dolls. Included were the all composition Nancy and Kewty dolls with molded hair and painted features. These two models were made to compete with Effanbee's more popular Patsy line. Many of the dolls were sold with trunks containing extra clothing. Most date from the early to mid-1930s.

In 1932, following the example of other doll companies, Arranbee dressed some of their regular inexpensive dolls in colonial costumes and marketed them as George and Martha Washington. This promotion marked the celebration of the 200th birthday of George Washington and the dolls came complete with "powdered" wigs.

The Arranbee company was also responsible for a series of all composition dolls in the 9" size. The same basic doll was used for a variety of characters, who achieved their individual identity from their costumes. Included were twenty different nursery rhyme figures, such as Jack and Jill, Little Red Riding Hood, Mary Had a Little Lamb, and Little Bo Peep. These same dolls were also dressed in costumes to represent the native dress of different countries.

For the most part, Arranbee dolls were of high quality but not very innovative—the company usually followed the examples set by other manufacturers. There were no Dy-Dee Babies, Shirley Temples, or other trend setting dolls developed by Arranbee. It is thought that some of the dolls sold by Arranbee were manufactured by other companies and only completed and dressed by Arranbee.

In 1947, the company began marketing dolls of hard plastic and continued to uphold their reputation for producing fine quality dolls. The firm was purchased by Vogue in 1958 and the Arranbee products were discontinued by 1960.

Arranbee (R & B)

11" My Dream Baby with original box reading "An R and B Quality Doll" The toddler style 1930s jointed doll has painted features, molded hair, and an open mouth. The doll is marked "Dream Baby" on the back. She is all original with her box bottom ($175).

10" all composition Arranbee "Drink 'n Babe" doll from 1935. She has curved arms and legs, painted features, molded hair, and an open mouth. She is marked on the back of her body, "Dream Baby." The doll is wearing her original clothing and still has her bottle, although the milk bottle is missing from the original box. The doll "drank" the milk in the bottle and the milk reappeared in the bottle when the bottle was placed upright ($175+).

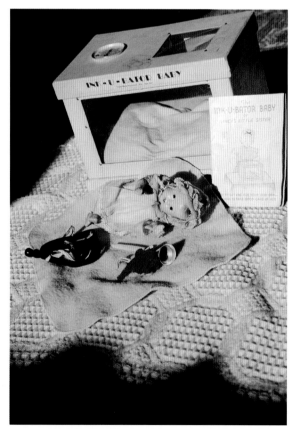

21" Arranbee doll, circa 1940, probably Little Angel Baby. She has a composition flange head with full composition arms and curved composition swinging legs. Her cloth body includes a crier box. She has a short lambskin wig, closed mouth, and sleep eyes with lashes. The doll is marked on the back of her neck "R & B." She is wearing her original dress, slip, shoes and socks. Her rubber panties have "melted" and been discarded. She also has a pink taffeta coat and bonnet which were factory made and may be original ($250).

Left:
6-1/2" composition Ink-U-Bator baby marketed by Arranbee in 1936. The jointed doll has painted eyes and molded hair. She is marked "R & B Doll Co." The doll is all original with her incubator box. Included in the set is a blanket, oxygen tube, stethoscope, water bottle (melted), pillow, and instructions. Missing is the nurse's cap that was in the original set ($250+). *Doll and photograph from the collection of Ellen Cahill.*

16" all composition Arranbee Nancy. She has sleep eyes, open mouth with teeth, and a real hair wig. She is marked on the back of her head "Nancy." She is wearing her original dress, underwear, and hooded coat modeled after the style so many little girls were wearing circa 1940s. This doll looks very much like the Alexander Princess Elizabeth dolls of the late 1930s ($225).

18" all composition Arranbee Nancy, circa the late 1930s. She has sleep eyes with lashes, open mouth with teeth, and a mohair wig. She is marked "Nancy" on the back of her head. She is all original, wearing a yellow organza dress, long slip and underwear, and a picture hat ($300+). *Doll from the collection of Jan Hersey.*

Left:
18" all composition Arranbee Nancy, circa late 1930s. She has sleep eyes with lashes, open mouth with teeth, and a real hair wig. She is marked "Nancy" on the back of her head. She is all original and near mint. Her original tag reads "Nancy/RandB/Quality Doll" ($400+). *Doll from the collection of Jan Hersey.*

Right:
20" composition Arranbee "Nancy the Movie Queen," circa 1939. She has sleep eyes, open mouth with teeth, mohair wig, and is fully jointed. She is all original wearing a pink organdy dress with petal shaped hem (not enough examples to determine price). *Doll from the collection of Marge Meisinger. Photograph by Carol Stover.*

Arranbee (R & B)

17-1/2" all composition Arranbee Nancy Lee, circa 1940s, dressed in skating costume. She has sleep eyes with lashes, a closed mouth, and a blonde mohair wig. These dolls were not meant to be skating star Sonja Henie because the Alexander Doll Co. was the only firm authorized to produce the authentic Sonja dolls. This doll is all original except for one missing ice skate blade. She also comes with her original box and tag. Her tag reads "Nancy Lee/An/R&B/Quality Doll." ($350+).

The original box for the Nancy Lee skating doll reads "R&B/QUALITY/DOLL/NANCY LEE/3116."

17" all composition Arranbee doll, probably Nancy Lee. She is very similar to the skating doll and is dressed in her original bride costume. She wears a taffeta ruffled underskirt beneath her thin rayon dress. She has sleep eyes with lashes, a closed mouth, and a mohair wig. She is marked "R&B" on the back of her head ($125+).

Left:
21" all composition Arranbee Nancy Lee, circa 1940s. She has sleep eyes, closed mouth, and a mohair wig. She is wearing her original clothing and retains her beautiful coloring ($400-500.) *Photograph and doll from the collection of Veronica Jochens.*

Far left and left:
21" Arranbee Nancy Lee in mint condition. She still has her original tag and box. She has sleep eyes with lashes, a closed mouth, and a mohair wig. The composition on the doll retains its beautiful color and the doll's hair has remained in its original set (not enough examples to determine a price). *Photograph and doll from the collection of Veronica Jochens.*

Left:
Rosalie teen-age doll, thought to be made by Arranbee, as advertised in the Sears Christmas catalog for 1938. The doll had a composition head, composition arms and legs, and a cloth body. She came in three sizes: 13-1/2", 18", and 24". The larger dolls had real hair wigs while the smallest doll came with a mohair wig. The dolls were priced from $1.98 to $4.98 each. *From the collection of Marge Meisinger.*

Right:
18" Arranbee teen-age doll, circa late 1930s, usually called Debu'Teen by collectors but may be the same model as Rosalie. Head, shoulders, arms, and legs are composition while the body is cloth. She has a swivel head on a composition shoulder plate. She has sleep eyes, closed mouth, and a real hair wig. She is dressed in her original yellow taffeta formal with net overskirt ($200).

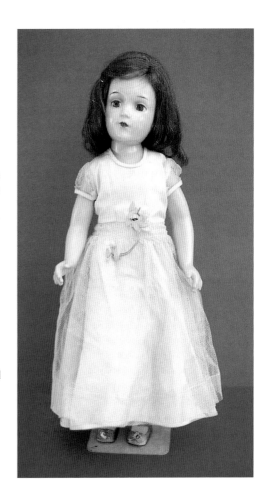

14-1/2" composition Arranbee doll, probably from the Southern Series. She has sleep eyes with lashes, closed mouth, and a real hair wig. She is marked "R&B" on the back of her head. She is all original and her flowered taffeta dress is trimmed with red ribbon. The wrist ribbons were probably once attached to her dress. She wears a hoop petticoat and long pantalettes. Her shoes have the cut-outs on the side often used by Arranbee ($225+).

12" all composition Arranbee Nancy, circa mid-1930s. She has painted eyes, closed mouth, and molded hair. She is marked on the back of her body "ARRANBEE." These dolls were marketed by Arranbee to compete with the Effanbee Patsy dolls. Her original wrist tag reads "NANCY/ARRANBEE/DOLLS." She still has her original suitcase with a complete wardrobe, which includes an organdy dress, hat, and one piece underwear, cotton dress, beach outfit, coat and hat, and pajamas ($200-250 complete).

Nancy's suitcase retains its original label identifying it as an Arranbee Nancy Wardrobe. None of the clothes are tagged.

14" composition Arranbee Nancy Lee with original tag. She has sleep eyes with lashes, closed mouth, and a real hair wig. Her tag reads "This Doll Has a Human Hair Wig." She still has her original box and a fur coat that is in tatters. Although she is all original, her thin rayon dress is in poor condition. She also has a pink taffeta ruffled underskirt and panties. Her shoes have the cut-outs on the sides used by Arranbee ($150+).

Arranbee also marketed a Kewty doll with a trunk full of clothes. The 14" all composition doll has sleep eyes, a closed mouth, and molded hair. She is marked on her back "KEWTY." This doll was also marketed with painted eyes and may have been sold as Nancy by Arranbee. She still has her original trunk and clothing. Cardboard hangers were included in the package. She is wearing her original velveteen coat and hat over an organdy dress. Her wardrobe also included a two-piece play outfit, sun suit, pique coat and hat, and pajamas ($250+).

9" all composition Arranbee dolls probably dressed in costumes representing a nursery rhyme—perhaps Jack and Jill. The dolls are identical, even though one is supposed to represent a boy. They have painted features, molded hair, and closed mouths. They are all original. They are marked on the back "R&B/DOLL CO." ($200 set).

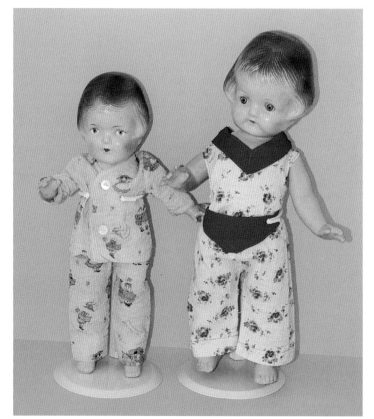

Nancy and Kewty model seersucker pajamas, which came in their trunks. Most of these trunk sets came with very inexpensive clothing except for one outfit that the doll was wearing.

Dewees Cochran

Dewees Cochran (1892-1991) was one of American's finest doll artists and her dolls continue to create interest for today's collector. Although Cochran was raised near Philadelphia, she lived, painted, and studied in Germany and Austria for nearly ten years. During this time, she met and married her husband, Paul Helbeck, a German writer. The couple returned to America after the death of Cochran's father in 1933. After a few years, Helbeck returned to Europe to pursue his own interests.

Dewees Cochran began making dolls soon after relocating to America in 1933. Her first efforts were black cloth dolls. These early designs included a topsy or turvy and a "mammy" doll. These character dolls would certainly be considered politically incorrect in today's world but in the early 1930s, the dolls sold well.

Acting on a suggestion from a doll buyer to make more realistic dolls, Dewees Cochran settled on a plan to produce portrait dolls that looked like their real life counterparts. She moved to New York City to allow her doll business to expand. One of her first commissions was to make portrait dolls representing the children of Irving Berlin. She made the doll heads of balsa wood and the bodies of kapok stuffed cloth.

While she was producing these higher priced portrait dolls ($85 each), Cochran was also researching the face types of real American children. She did this by studying pictures from library collections and photographs from model agencies. Cochran came up with six different types of faces that she began using as her models to make doll heads of plastic wood.

In 1935, Dewees Cochran was contacted by the Alexander Doll Co. and was offered a job with the firm. Cochran accepted with deep reservations and worked for only a short time before deciding that she would rather produce her own dolls than work at the Alexander factory.

By late 1935, the Dewees Cochran dolls were so well known that Cochran was approached by another famous doll company, Fleischaker and Baum (Effanbee Doll Co.), to work for that firm. The artist signed a three year contract to design dolls, based on her ideas, that looked like real children.

Relying on her earlier research, Cochran made six dolls of plastic wood. Effanbee decided to use four of these examples as models to produce dolls for their American Children series. The models, provided by Cochran, were 21" tall with painted eyes and human hair wigs. The dolls had individual fingers on their hands so they could wear gloves—a first in American doll production. The dolls were supposed to represent eight-year-old little girls and the first dolls in the series were available to the public in 1936. The arms and hands of the dolls were made of a latex or hard rubber material while the rest was composition. In 1938, a boy doll was produced, also with painted eyes.

Also in 1938 (or perhaps as early as 1937; Cochran's autobiography gives conflicting dates), Effanbee asked Cochran to design an open mouth smiling doll to represent a five or six-year-old child. This doll was to be 15" tall but was eventually produced in several sizes. These newer dolls had sleep eyes and Cochran had to adjust her design so the machine that set the eyes could be used on the dolls. According to her autobiography, Cochran also had to adjust the neck sizes as well as the eye sockets on the earlier heads to allow compatibility with the eye setting machine. This made it possible for F&B to produce the earlier line of dolls with sleep eyes. To allow for this process, some of the faces also had to be filled out. These alterations took away some of the individuality that was shown in the four different designs of the original 21" painted eye dolls.

The most common dolls in the American Children series are the later dolls with open mouths. These dolls were made in 15", 17", and 21" sizes. A tag came on the 15" doll identifying it as Barbara Joan. This doll was also sold as Sonja On Skates in the N. Shure Co. catalog in 1938, and, wearing proper costumes, as Ann Shirley and Snow White. The 17" dolls were tagged Barbara Ann and the 21" dolls as Barbara Lou. All of these dolls are marked on their backs "Effanbee/Anne Shirley." Their original tags identify the dolls as "One of America's Children."

Only the dolls with closed mouths are marked "Effanbee/American/Children" on the backs of their heads. Their bodies are marked "Effanbee/Anne Shirley." The early dolls in the series came in boxes marked "Portrait Doll/One Of America's Children." Some of the 21" dolls with closed mouths and sleep eyes were tagged Peggy Lou. The closed mouth dolls are the most desirable dolls for collectors, especially those with painted eyes. The most prized dolls of all are the painted eye boy dolls because not so many of these dolls were produced.

The Effanbee Co. was so pleased with the dolls that they used the painted eye, closed mouth dolls to model elaborately made clothing representing a history of American fashion. Thirty different costumes were displayed on these dolls. Three sets of the dolls were produced for an Effanbee promotion and they toured the country in 1939 in order to publicize the firm's new line of 15" Historical Dolls to be sold to the public.

Although Dewees Cochran had always received good publicity for her skill in doll production, the cover story of *Life* magazine's April 3, 1939 issue was probably her best. Writing in her autobiography, Cochran says she was approached by writers from *Life* to do a story on her famous portrait dolls. She told them that she felt the story would be bigger if they used the dolls being sold by Effanbee and tied a promotion from the Saks Fifth Ave. store into the story as well. Cochran contacted Saks and arranged for an artist to paint the features and complexions of the dolls to match their future owners for "Look-Alike" dolls. Dresses in the children's department were copied to be worn by the dolls (hopefully, the customer would also purchase a matching child's dress). Cochran arranged to use four child models for the *Life* story. These little girls represented the four different types of children's faces portrayed in the Effanbee American Children series. Four dolls were dressed to match the young model's clothing and each doll had the same hairstyle and coloring as its model. These were the little girls and dolls photographed for the *Life* magazine article. One of the models and her doll appeared on the cover. All of the dolls featured in the magazine article had sleep eyes and closed mouths. The Saks Fifth Avenue store promotion lasted for one week in April 1939 and the personalized dolls sold for $25 each.

Cochran had continued to make her own portrait dolls throughout her contract years with Effanbee. Because of the large number of orders, she secured extra help with costuming and sculpture to help meet the demand. During the World War II years, it was impossible to obtain materials to make dolls and Cochran turned to other work. At the same time, she continued to experiment with new materials to make dolls.

In 1947, after the war ended, the Dewees Cochran Doll, Inc. company was formed in order to develop a new 16" doll called Cindy. Dewees Cochran partnered with two small companies to produce the dolls. One thousand were made from a material called "Kaysam," which was very similar to the Vultex product Cochran used for her individual portrait dolls. The Cindy dolls were marked "Dewees Cochran Dolls" on the left side of the torso. According to Cochran, "Only these dolls are true to the original design." The other factory's Cindy dolls did not meet the designer's expectations and she withdrew from the arrangement.

A brochure for the acceptable dolls included pictures of several costumes including: 1. Playsuit, 2. All-Day Dress, 3.Town and Country Coat, 4. The Good Suit, 5. School Dress and Pinafore, 6. The Party Dress. The Cindy doll was featured in color in *Harper's Bazaar* magazine in 1947. The dolls could be purchased alone or with several changes of clothing in a shadow box carrier.

In 1952, Dewees Cochran began making a series of dolls out of Vultex called the Grow-Up Dolls. The line included a red-haired doll called Susan Stormalong or Stormie, a blonde wigged girl named Angela Appleseed, and a brunette called Belinda Bunyan. The dolls began life as three to five-year-olds but, as time passed, the dolls were reissued in dolls representing seven, eleven, sixteen, and twenty-year old youngsters. In 1954, boy dolls called Peter Ponsett and Jefferson Jones were added to the line. These dolls were eventually produced representing boys of five, fourteen, and twenty-three years of age.

Cochran moved to California in 1960, where she continued to make her dolls until she was in her eighties. In 1976, she was invited by Effanbee to design a doll for their limited edition series. Cochran decided to use a self portrait doll she had created for the United Federation of Doll Clubs Convention earlier in the decade. She redesigned the doll and costume, which portrayed the artist as a child of eight, and it became Effanbee's limited edition doll for 1977. The edition of approximately three thousand dolls sold out quickly. In 1979, Cochran's autobiography, *As If They Might Speak*, was published by Paperweight Press in Santa Cruz, California.

In 1980, the Dewees Cochran Foundation was founded in Orwell, Vermont. The foundation was located in a Federalist home in Orwell and housed Cochran dolls, sketches, writings, and other memorabilia. In 1984, the foundation, through the Dewees Cochran Doll, Inc. company, began issuing dolls made from the original Cochran molds. Several different dolls were marketed, including the Grow-Ups and Cindy. Belinda, Susan, and Angela were issued in 10" sizes. The dolls were made of porcelain and one hundred were sold at $535 per doll. These dolls were from Su'ben's Art, Orwell, Vermont. In 1988, the museum was selling Cindy dolls made of a composition type material as well as a Russian child made of porcelain.

Dewees Cochran passed away in 1991 at the age of ninety-nine. She spent the last years of her life in a nursing home in Santa Cruz, California.

Even though Dewees Cochran is gone, her dolls never cease to bring pleasure to new and old collectors alike. Effanbee continues to issue replicas of Cochran dolls based on her original designs. These include limited editions of Cindy and examples from the Grow-Up series. Although collectors of the original Cochran dolls will find that the dolls continue to escalate in price, most fans want to include at least one Dewees Cochran doll in their doll families.

21" Effanbee "American Children" doll, designed by Dewees Cochran circa 1936-37. She has painted eyes, closed mouth, and a human hair wig. Her arms and hands are made of a latex material while her head and body are composition. The hands on these dolls, which were supposed to represent eight-year-old children, have individual fingers so the dolls are able to wear gloves. The doll is wearing her original clothing and still has her Effanbee bracelet. She is marked "Effanbee/American/Children" on the back of her head and "Effanbee/Anne-Shirley" on her back ($1,250-1,500). *Doll from the collection of Lois Jakubowski. Photograph by Robert Jakubowski.*

These 21" dolls, with the painted eyes, were issued in four different models, each featuring a different shaped face as designed by Dewees Cochran. *Doll from the collection of Lois Jakubowski. Photograph by Robert Jakubowski.*

Life magazine did a cover story on the Effanbee "American Children" dolls in their April 3, 1939 issue. The story featured the four different shaped faces of the dolls and matched them to four real children. All of the dolls used in the story were the later dolls from 1939 which had closed mouths and sleep eyes ($50).

17" Effanbee "American Children" Birthday Doll, as pictured in the 1939 Montgomery Ward's Christmas catalog. She has almond shaped sleep eyes, closed mouth, and a real hair wig. She is all composition except for her arms and hands, which are a hard rubber or latex material. The "Happy Birthday" windup music box can be activated from her back. She is marked "American Children" on the back of her head and "Effanbee/Anne-Shirley" on her back. The doll is all original, complete with her metal Effanbee heart bracelet ($1,000+).

Effanbee "American Children" dolls were featured in the Montgomery Ward Christmas catalog in 1939. The magazine publicity was mentioned although the publication name was omitted. Both dolls had sleep eyes and closed mouths. The 20" model sold for $4.95. The smaller 17" doll sold for $6.95 and included a new innovation, a windup music box that played "Happy Birthday." The text says that both dolls have hard rubber arms with the rest of the dolls being made of composition. *From the collection of Marge Meisinger.*

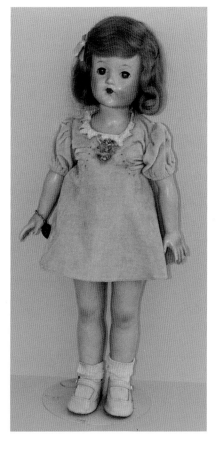

20" Effanbee "American Children" doll, as shown in the 1939 Montgomery Ward Christmas catalog. She has more rounded sleep eyes, a small closed mouth, and a real hair wig. She is all composition except for her arms and hands, which are a hard rubber or latex material. She is marked "American Children" on the back of her head and "Effanbee/Anne-Shirley" on her back. The doll is all original and still wears her Effanbee metal heart bracelet. The birthday doll and this doll illustrate two of the four different faces pictured in the *Life* magazine article ($700+).

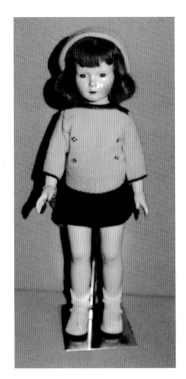

20" Effanbee "American Children" doll, circa 1939. She has almond shaped sleep eyes, closed mouth, and a real hair wig. She is all composition except for her arms and hands, which are hard rubber or a latex material. She is marked "Effanbee/American Children" on the back of her head and "Effanbee/Anne-Shirley" on her back. She is all original and still wears her Effanbee metal heart bracelet ($900+). *Doll from the collection of Lois Jakubowski. Photograph by Robert Jakubowski.*

21" Effanbee "American Children" doll, circa 1939. Although some of the dolls are 20" and many are 21", it is likely that the way they were molded caused the difference in size. This doll is wearing her original clothing and is marked in the same manner as the other dolls. Her mouth is very tiny and her eyes are more rounded, like the doll featured in the Montgomery Ward 1939 catalog ($700+). *Doll from the collection of Lois Jakubowski. Photograph by Robert Jakubowski.*

20" Effanbee American Children doll, circa 1939. She is wearing her original coat, hat, and gloves. Since these dolls had hands with fingers molded individually, gloves were often used as part of their costumes. Her markings and descriptions are the same as the other dolls in the closed mouth series ($800-900). *Doll from the collection of Lois Jakubowski. Photograph by Robert Jakubowski.*

Advertisement from the John Plain catalog from 1940 for an Effanbee American Children 21" Portrait Doll. The early closed mouth dolls came in boxes marked "Portrait Doll/One of America's Children." The doll has sleep eyes, closed mouth, and a human hair wig. *From the collection of Marge Meisinger.*

The face on this "American Children" doll appears to be a little fuller than the face on the doll in the yellow sweater. Since Cochran had to modify her original designs to accommodate for the sleep eyes, it makes it harder to identify the four different faces. *Doll from the collection of Lois Jakubowski. Photograph by Robert Jakubowski.*

PORTRAIT DOLL
A gorgeous portrait doll of highest quality. She stands 21 in. high. You can comb and curl her golden human hair. Dressed in the height of fashion, a Paris-inspired dress with jacket; ascot tie around neck; ribbon trimmed bonnet. Sleeping eyes with real lashes. Fully jointed body.
N26457 PRICE..........$12.50

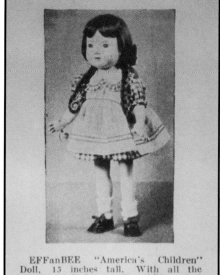

EFFanBEE "America's Children" Doll, 15 inches tall. With all the quality and charm with which a doll can be endowed. Typical EFFanBEE exclusive features such as manicured finger nails, self standing, life-like smooth finish, almost human glass eyes, lashes, human hair wig in braids, and many others. All composition body with jointed arms and legs, and turning head. Dressed in gingham dress with matching combination underwear. Organdy pinafore trimmed with braiding to match the color of the dress. Ribbon bow tied in hair. Shoes to match outfit; white silk socks.
No. 36N317. Each.........$5.80

N. Shure Co. 1938 advertisement for a 15" open mouthed doll designed by Dewees Cochran. Although these dolls are not marked "American Children," tags and advertisements sometimes referred to them in this manner. The design for this line of dolls was supposed to represent a five-year-old child. *From the collection of Marge Meisinger.*

21" Effanbee American Children doll with original Peggy Lou tag. She is wearing the same shoes as shown on an American Children doll in a 1940 catalog. She is all original, complete with her gloves and Effanbee metal heart bracelet. She has the almond sleep eyes, a closed mouth, and a real hair wig. Her tag reads "I Am/Peggy Lou/An EFFanBEE/DURABLE/DOLL/One of America's Children." Her coat and hat are made of wool and she wears a red print cotton dress. Pictured with her are two 15" Effanbee dolls with open mouths and sleep eyes. They wear their original clothing. They are from Dewees Cochran's open-mouthed series of dolls circa 1939 (21" sold at auction in 1999 for $2,500; smaller dolls at same auction: top right $350, bottom right $650—buyer's premium not included.) *Photograph courtesy of Frashers' Doll Auctions, Inc.*

Left:
17" Effanbee composition Barbara Ann from the open mouth series of dolls designed by Dewees Cochran. All of these dolls are marked "Effanbee-Anne Shirley" on the torso. Her tag reads "I am one of/AMERICA'S CHILDREN/Barbara/Ann/AN EFFanBEE/PLAY/PRODUCT/MADE IN/USA." The doll is all original, including her Effanbee metal bracelet ($600).

Right:
15" all composition Barbara Joan doll as pictured in the N. Shure ad from 1938. She has sleep eyes, open mouth with teeth, and a human hair wig. The ad mentioned her manicured fingernails. She is all original ($600-700). *Photograph and doll from the collection of Jo Barckley.*

N. Shure Co. 1938 advertisement for a 17" doll from the open mouth series. She was described as being made of composition with glass eyes, eyelashes, human hair page boy wig with bangs, open mouth with teeth, and manicured fingernails. *From the collection of Marge Meisinger.*

21" Effanbee open mouthed Barbara Lou doll. She has sleep eyes, open mouth with teeth, and a human hair wig. She is marked "Effanbee/Anne-Shirley" on her torso. She is all original, wearing a cotton print dress with a stole ($700-800). *Doll from the collection of Lois Jakubowski. Photograph by Robert Jakubowski.*

21" Effanbee composition Barbara Lou from the open mouth series of dolls designed by Dewees Cochran. She still has her original tag reading "I am one of/AMERICA'S CHILDREN/BARBARA LOU." She is all original and near mint condition. She has sleep eyes with lashes, an open mouth with teeth, a human hair wig, and is wearing her original Effanbee metal bracelet ($800-850). *Doll from the collection of Lois Jakubowski. Photograph by Robert Jakubowski.*

The face of the 21" open mouthed Barbara Lou doll still retains its original color. These dolls are all marked "Effanbee/Anne Shirley" on the torso. *Doll from the collection of Lois Jakubowski. Photograph by Robert Jakubowski.*

N. Shure Co. 1938 advertisement for the 21" America's Children open mouth doll. The ad calls this doll the company's "finest doll." The copy describes her as being all composition, fully jointed, with moving eyes, and a human hair wig. She was wearing a flannel silk lined coat with Scotch Plaid cuffs, lapels, and buttons, and matching Scotch Plaid hat with a feather. *From the collection of Marge Meisinger.*

This trunk full of clothes contains several outfits made for the American Children line of dolls. *From the collection of Lois Jakubowski. Photograph by Robert Jakubowski.*

Coat and gloves often seen on the 20"-21" Peggy Lou doll (sleep eyes and closed mouth). The costume originally included a hat ($100-150). *From the collection of Lois Jakubowski. Photograph by Robert Jakubowski.*

Since the clothes in the trunk are not marked, it is difficult to determine if they were actually sold with the dolls or if they were part of a promotion. It is known that the clothes sold with the Dewees Cochran/*Life*/Saks Fifth Ave. special event were made especially for that promotion. Shown here is a red skirt, matching jacket, and blouse made to fit the 21" dolls. *From the collection of Lois Jakubowski. Photograph by Robert Jakubowski.*

Cotton flowered summer formal along with a short slip. *From the collection of Lois Jakubowski. Photograph by Robert Jakubowski.*

Dewees Cochran 45

Red and white striped cotton dress, belt, and matching panties, also included in the trunk. *From the collection of Lois Jakubowski. Photograph by Robert Jakubowski.*

The "American Children" doll who owned these clothes had her choice of a cotton flowered robe for summer or a long sleeved yellow robe for winter. *From the collection of Lois Jakubowski. Photograph by Robert Jakubowski.*

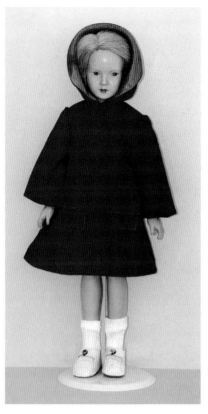

15" Dewees Cochran Cindy doll, circa 1948. She is made of a latex material called Kaysam. The fully jointed doll has painted eyes, closed mouth, and a real hair wig. She is wearing her original "Town and Country Coat." Only one thousand Cochran approved dolls were sold. They are marked "Dewees Cochran Dolls" on the left side of the torso; this one is numbered 879. These dolls were sold by the Dewees Cochran Doll, Inc. company ($400-500 with moth holes in coat and possibly replaced shoes).

Whitman Publishing Co. produced a set of paper dolls in 1939 called Patsy Lou and Barbara Lou #976. These dolls were based on the 21" Effanbee Barbara Lou from the open mouth American Children series and the 22" Patsy Lou from the Patsy line of dolls. Pictured is the Barbara Lou doll with some of the clothes shown in the book. The formal is a replica of the dress worn by the Barbara Lou doll pictured earlier. The dress and hat have also been seen on real Barbara Lou dolls.

Effanbee Doll Corporation

The Effanbee Doll Corporation, based in New York City, had its beginning in 1910 when Bernard E. Fleischaker and Hugo Baum joined forces to sell toys. In 1912, the firm began to manufacture dolls. Their trademark became Effanbee in honor of the founders initials: F & B. The firm made many successful composition dolls in the 1920s, including Baby Dainty, Bubbles, Pat-O-Pat, Baby Grumpy, Mae Starr, and the first dolls from the Patsy family.

The earliest Effanbee dolls carrying the "Patsy" name were made circa 1927. These dolls really had no relation to the later Patsy family dolls. The 1927 dolls had composition shoulder, heads, and full arms and legs, along with cloth bodies, sleep eyes, and open mouths with teeth. Each doll was marked "EFFANBEE/PATSY" in a half circle on the back.

As the 1930s began, other Effanbee dolls being produced included "Mama" dolls of various kinds with composition heads and lower limbs, and cloth bodies. Most of these dolls had sleep eyes and wigs and were dressed as little girls instead of babies. A more unusual Effanbee product was the Lampkin baby doll. This doll had very curved composition arms and legs, a composition head, and cloth body. The ads called the doll "Patsy's new baby sister" but the doll was larger than the Patsy dolls. Bubbles dolls that were such a success in the 1920s were also still being marketed as the 1930s began. By 1932, the dolls were being made with rubber bodies.

Much of the publicity for Effanbee dolls in the early 1930s revolved around the ever-growing Patsy family dolls. The phenomenon began with the introduction of a doll called "Mimi" in 1927. The all composition doll was designed by famous doll designer, Bernard Lipfert. The doll artist wanted the doll to represent the look of a real child. Although advertising from December 1927 called this new doll "Mimi," by early 1928 she had been renamed Patsy and was being marketed under the new title. Some dolls marked "Mimi" were sold but they are rare. By 1936, all of the various sizes of the Patsy dolls had been introduced.

The following dolls, listed by size, were included in this family of dolls. Approximate date of introduction is also included:

6" Wee Patsy (1932)
7" Baby Tinyette (1933)
8" Patsy Tinyette (1933)
8" Patsy Babyette (1933)
9" Patsyette (1931
10" Patsy Baby-kin (1932)
11" Patsy Baby (1932)
11" Patsy Jr.- Patsykins (1930)
11" Patricia-kin (1932)
14" Patsy (1928)
15" Patricia (1932)
16" Patsy Joan (1931)
19" Patsy Ann (1929)
22" Patsy Lou (1931)
26" Patsy Ruth (1936)
30" Patsy Mae (1934)

The Patsy dolls were offered in a variety of costumes from year to year. Some dolls were marketed with trunks or in special boxes that contained extra clothing. Others were, at one time, fitted with magnets in their hands. These included some of the Patsy, Patsy Joan, Patricia, and Babykin dolls. Patsy Joan was also sold with a music box in her body in 1940. Some of the dolls were produced with brown skin as well as in flesh color. Included were Patsy, Patsy Baby, Patsyette, Patsy Ann, Patsykins, Patricia, and Patsy Joan. Patsy was also designed as a Chinese doll with a different style painted black hair. She is very rare. The Effanbee firm followed the same practice as Alexander by marketing their Patsy family dolls in pairs. These dolls were supposed to represent both girl and boy characters. The only difference in the dolls was their clothing, with the boy dolls dressed in male outfits and the girl dolls wearing female attire.

The original Patsy dolls came with painted eyes and molded hair. As the family grew, sleep eyes were added to the larger dolls. Wigs replaced the painted hair on many of the dolls by 1933.

Several of the Patsy family dolls were used to represent Anne Shirley, the main character in the 1934 movie, *Anne of Green Gables*. Anne was played by an actress who then took the name as her own and she became known as Anne Shirley (see Personality Dolls chapter). Dolls used to tie in to this film included Patricia, Patricia-kin, and Patsyette. These dolls all had wigs styled in pigtails.

The Patsy dolls were so popular that additional clothing was made just for these dolls. Several different firms provided the outfits but the most well-known is the Glad Togs company headed by Gladys Myers in Seattle, Washington. Many of the "Glad Togs" tags still remain on this early clothing. Several different Patsy family commercial doll clothes patterns were also sold by many of the large pattern companies. Included were McCall, Butterick, and Simplicity (see Pattern and Doll Clothing chapter).

Besides doll clothing and trunks, the Patsy family inspired other products that were made to capitalize on their images. In an EFFanBEE ad from December 1931, the firm urged little girls to enroll their doll in the "New Patsy Doll Club." She would receive a membership certificate, a pin, and the new magazine. "Aunt Patsy" played a role in this publicity when she traveled around the country to large department stores to promote the Patsy dolls. This character also told stories about the dolls in the company publication. A book called *Patsy for Keeps* was published by Sam. Gabriel Sons & Co. in 1932. This was a "Patsy and Patsy Ann Put Together Book." The owner was to illustrate the book by cutting out the pictures from gummed sheets and placing them in the proper places in the book.

Another children's book, *Patsy Ann: Her Happy Times*, by Mona Reed King was published by Rand McNally & Co. in 1935. This book was illustrated with photographs of Patsy Ann wearing different outfits and participating in various activities. One of the most collectible of the Effanbee "extras" was the paper doll book "Patsy Lou Barbara Lou" #976, published by Whitman Publishing Co. in 1939. The two large paper dolls represented Patsy Lou from the Patsy family of dolls and Barbara Lou from the Dewees Cochran designed American Children doll series.

The Patsy family dolls remained popular through 1940, although most of their success occurred from 1929-36. In 1946, the composition Patsy and Patsy Joan dolls were reissued and remained on the market through 1949. The Patsy Joan dolls were sold with brown skin as well as flesh color. The "new" Patsy was featured in the Montgomery Ward Christmas catalog for 1946.

Although Skippy made his first appearance in 1929, he remained a part of the Effanbee doll family for many years. He was based on a cartoon character drawn by Percy L. Crosby (see Character and Military Chapters). The Skippy doll, dressed in the costume from the Jackie Cooper Paramount *Skippy* movie, dates from 1931. Brown Skippy dolls were also marketed at a later time.

The next big success in the Effanbee doll world was the introduction of the Bernard Lipfert designed Dy-Dee baby doll, in 1934. This doll had a hard rubber head and a soft rubber jointed body. The earliest dolls had molded ears, molded painted hair, and sleep eyes. They could be purchased with or without layettes and the layettes were also sold separately. Dy-Dee dolls were made with a "drink and wet" design and were offered in a variety of sizes, including 9", 11", 13", 15", and 20". These Effanbee best sellers were featured in wholesale and mail order catalogs year after year. Besides layettes, Dy-Dee Bathinettes were available by 1938 to use for bathing the rubber dolls. Other tie-in products included a "Mother Outfit" featured by the John Plain catalog in 1938. The suitcase set included a rubber apron, bottle, powder, powder puff, washcloth, hot water bottle, lawn uniform and cap (like nurse), and a Dy-dee Diary and pencil.

The Dy-Dee Baby doll ads in the N. Shure catalog for 1941-42 pictured the "new" Dy-Dee baby and the text said she was new for 1941. This doll had soft flexible ears along with "real pink-lined nostrils." She could drink from a bottle, sip from a spoon, or blow bubbles. She retained a hard rubber head and soft rubber body. This new model was listed in sizes of 11", 15", and 20". A packaged Dy-Dee-Ette was also available with a layette. She came in the same sizes. A Deluxe Dy-Dee Louise was also pictured, in 11", 15" and 20" models. She and her larger layette were packaged in a "Luxurious Alligator Grained Luggage Case."

By 1942, the dolls were featured in the Montgomery Ward Christmas catalog with lambskin wigs. In the 1948 Montgomery Ward Christmas catalog, it was announced that a crier-pacifier had been added to the Dy-Dee dolls that year. The same company featured Dy-Dee dolls with either molded hair or lambskin wigs in 1949. By that time, the dolls had heads made of hard plastic while they retained their rubber bodies.

More products were made to tie in to the Dy-Dee dolls. The *Dy-Dee Doll's Days* book, by Peggy Vandegriff, was published by Rand McNally in 1937. It was illustrated with photographs of the Dy-Dee doll in everyday activities. A paper doll book was also published featuring a paper doll in the image of the Dy-Dee doll. The "Dy-Dee Baby Doll" paper dolls #969 were published by Whitman Publishing Co. in 1938. A pattern called "Baby Clothes for Dy-Dee Dolls" #632 was issued by McCall in 1938 and continued to be sold for many years (see Patterns chapter).

Other Effanbee dolls marketed during the 1930s included the "Tousle Head" dolls. These dolls came in a variety of sizes and styles but all featured the fur-lambskin wigs. Some of the dolls were all composition little girl models and others had cloth bodies and baby styled arms and legs. The Montgomery Ward Christmas catalog for 1939 featured one of these dolls, which carried a tiny phonograph in her body that played "Now I Lay Me Down to Sleep." The doll was 18" tall, with a composition head, arms and legs, soft body, and sleep eyes. She was priced at $12.95.

While other major doll companies of the 1930s were marketing personality dolls, Effanbee did not produce a famous Shirley Temple doll or authorized Dionne Quint products. The company did, however, have famous doll artist Dewees Cochran. The wonderful dolls produced by the company from Cochran's models are some of the most collectible composition dolls ever made (see Dewees Cochran chapter). In 1939, Effanbee manufactured a set of thirty composition dolls (with hard rubber arms) representing the "Fashion History of America." The dolls were made using Cochran's 20"-21" closed mouth, painted eye models from the "American Children" series. They were beautifully dressed in historical costumes and were exhibited in large department stores around the country. At the same time, the firm marketed 15" composition dolls wearing the same types of costumes (only made of cotton). These dolls were sold to the public. There were thirty dolls in this series also, referred to by collectors as replica Historical Dolls. They were made using the Little Lady heads and Anne Shirley bodies. The dolls had painted eyes, closed mouths, and wigs (see Polly and Pam Judd's book *Compo Dolls 1928-1955* for a complete list of these dolls).

Effanbee did market a couple of successful personality dolls in 1938 when they produced W.C. Fields and Charlie McCarthy ventriloquist type dolls. (see Personality section). A less known doll is Sonja on Skates, also from 1938. Although the doll was not an authorized Sonja Henie product, it did tie in to the ice skater's successful career. It was advertised in the 1938 N. Shure Co. catalog in a 16" size. The doll had a human hair wig, sleep eyes, and used Dewees Cochran's open mouth child doll design. The doll was also called "Ice Queen."

The Effanbee Lovums dolls were also sold during the late 1930s and early 1940s. These dolls were made with either molded hair or lambskin wigs. They had composition heads and shoulder plates and full composition arms and legs to above the knee, with cloth bodies. These baby dolls had sleep eyes and open mouths with teeth. The N. Shure Co. catalog for 1941-42 pictured one of these dolls in a new "Heartbeat" model. It came with a stethoscope so its owner could hear the doll's heartbeat. The doll was listed in sizes of 17" and 19".

The Anne Shirley dolls made in the last part of the 1930s had their own mold and were not Patsy family dolls dressed in *Anne of Green Gables* clothing. These dolls were marketed in several different sizes and costumes (see Personality Dolls chapter). By 1939, the dolls with Anne Shirley bodies were being sold under the "Little Lady" trade name. They continued to be popular until the end of the 1940s. The more recent dolls were not marked Anne Shirley on the back but, instead, were identified with the Effanbee name. The Little Lady dolls were frequently advertised in mail order catalogs in sizes that included 15", 18", 19", 20", 22", and 28". The composition dolls had sleep eyes, closed mouths, wigs, and were fully jointed. They were dressed in a variety of costumes including long gowns, negligees, bridal attire, majorettes, and little girl dresses. A few of the dolls came with wardrobes and, in 1940, Montgomery Ward carried a Little Lady with magnets in her hands. Due to World War II and the scarcity of materials, many of the Little Lady dolls were made with painted eyes and heavy thread-type hair from 1943-45. Some of these dolls had cloth bodies and composition heads and hands only. The popularity of the Little Lady dolls also influenced the production of tie-in products. Included were a puzzle called "Little Lady Jig Saw Puzzle" with a 1943 copyright by Fleischaker & Baum and a doll clothes pattern by McCall. The pattern (#1089) states that "the clothes will fit the Little Lady dolls." It carried a copyright of 1943 from the McCall Corporation (see Pattern chapter). The puzzle picture came from the two-page ad used in the Montgomery Ward Christmas catalog in 1943.

Along with the Patsy family and Dy-Dee dolls, the Little Lady models were Effanbee's most long lasting composition dolls.

In addition to the small Patsy family dolls, Effanbee offered other smaller, more inexpensive dolls in the 1930s and 1940s. One of these was "Butin-Nose" (correct spelling from Ursula R. Mertz's book *Collector's Encyclopedia of American Composition Dolls 1900-1950*). These 8" composition dolls were dressed in a variety of costumes, including those representing native dress from foreign countries, and as ordinary little girls and boys. The Chinese doll is especially unusual because of the hair painting and costume. All of the dolls had painted features, molded hair, and were fully jointed.

Little girl dolls were also produced in smaller sizes. These included an unidentified 9" all composition doll with molded hair and painted features that looked very much like the 11-1/2" marked Suzette dolls. These dolls, with molded hair, sold for under $1.00 in 1940. The molded hair Suzette doll was pictured in the Montgomery Ward Christmas catalog of that year priced at 95 cents. The most collectible Suzette dolls are those dressed to represent George and Martha Washington. The 9" Patsyette dolls were also sold in George and Martha Washington costumes. It is assumed that these dolls were marketed circa 1932 during the Bicentennial celebration of George Washington's birth. All of these dolls had painted features but "powdered" wigs were added. Some of the Suzette little girl dolls also came with wigs. Other very desirable Suzette dolls include those made with brown skin, which were dressed in authentic Hawaiian costumes. These dolls had molded hair.

The Effanbee 14" Suzanne dolls were also popular during the late 1930s and early 1940s. These dolls were marked "Suzanne" on their heads. They came with sleep eyes and wigs. Some were packaged in suitcases with extra clothing and a few had magnetic hands. One of the suitcase sets included nurse's outfits and accessories. This unit was sold by F.A.O. Schwarz. In 1942, Suzanne was also marketed dressed in W.A.A.C and W.A.V.E uniforms. Due to World War II and the scarcity of materials, some of these dolls had painted eyes.

The 12" Effanbee composition Portrait dolls joined the Effanbee family of dolls circa 1942. These dolls had sleep eyes, closed mouths, mohair wigs, and were fully jointed. The dolls were unmarked so collectors need to be familiar with the looks and costumes of the dolls in order to identify them. Some apparently had dress tags, which has helped in their identification. Effanbee made some of these dolls in pairs and, following their usual practice, they used the same doll for both male and female characters. Only the hairstyles and the clothing changed the dolls from female to male figures. Several different characters have been identified, including a bride and groom, Spanish Dancers (or just Dancers), Gibson Girl, Ballerina, and Little Bo Peep.

As World War II intensified in 1943, doll materials became scarce and manufacturers had to improvise in order to keep marketing their products. They sold dolls already in their inventories, then began marketing dolls made mostly of cloth. These war-time dolls were usually produced with composition heads and hands and cloth bodies and limbs. Even the mohair wigs and sleep eyes used materials needed for essential products. The new dolls featured painted eyes and thread-yarn hair to avoid the shortages. A pair of Effanbee Brother and Sister dolls was marketed in 1943 using these new guidelines; the author received a set of these dolls for Christmas that year. The Montgomery Ward Christmas catalog devoted a full page in color to a picture of these special dolls. The Brother was 16" tall and the Sister measured 12". The pair was priced at $6.95 in the catalog.

The Effanbee "Babyette Asleep" doll pictured in the Montgomery Ward catalog for 1945 followed some of these same principles of war-time production. Although the war ended in August of 1945, it took some time for toy manufacturers to return to normal. This baby doll was 12" long and came with a composition head and hands, a cloth body and limbs, molded hair, and "eyes painted shut." Montgomery Ward offered the doll, sleeping in a basket, for $16.95.

The same catalog also pictured some Effanbee baby dolls with sleep eyes and lambskin wigs. Included was a "Sweetie Pie" with flirty eyes and a cry box. This doll had a composition head, composition arms and legs, and a cloth body. She was listed in sizes of 19" and 24". The "Sweetie Pie" head was the same one used on Effanbee dolls called Tommy Tucker, Mickey, Bright Eyes, and the Mickey and Janie Twin dolls. Some of the dolls had molded hair, some had mohair wigs, and others had the lambskin wigs. These dolls, in one form or another, remained on the market for most of the 1940s. The author received a doll that collectors now call Mickey for Christmas in 1944. At that time he may have been called "Sweetie Pie." He was dressed in corduroy overalls, long sleeved shirt, and beanie hat just like the Mickey dolls. His construction was probably a result of war-time shortages as only his head and hands were composition; the rest of the doll was cloth. He did have a mohair wig and sleep eyes as did many of the later Mickey dolls. Both the Mickey and Sweetie Pie dolls were still being advertised in the *Children's Activities* magazine in 1947. The Baby Bright-Eyes (16") and the Mickey and Janie twins (14") were marketed in suitcases with extra clothes circa 1946. All of these dolls had wigs.

One of the last of the unique composition Effanbee dolls to be marketed was the Candy Kid doll, introduced circa 1946. These all composition dolls were 13" tall and had sleep eyes, closed mouths, molded hair, and were fully jointed. They came dressed as either boys or girls and as a pair, which included a girl and a boy in the set. One of the most popular outfits for today's collector is Candy Kid's boxing costume, which came complete with boxing gloves. This doll was made in both brown and flesh colored styles. Some of the Candy Kids came with suitcases and extra clothing. The 1946 Montgomery Ward catalog offered such a set for $16.50.

The Effanbee company was sold to Noma Electric Corp. in 1946 and the new company developed some vinyl dolls (including Lil Darlin) as well as a composition Howdy Doody. By 1948, the composition Honey doll was in production. This doll would become a hit when it was made of hard plastic in the 1950s. By 1950, Effanbee had joined the band wagon and nearly all of its dolls were made of hard plastic.

In 1953, Bernard Baum (son of original founder), Perry Epstein, and Morris Lutz bought the company and assumed ownership. Despite changes in management, Effanbee produced many excellent examples of collectible dolls during the decade of the 1950s. Perhaps the most attractive to today's collectors are the mint condition hard plastic Honey dolls, when found with their original crisp clothing and excellent skin coloring. These dolls rank at the top of any hard plastic doll collection.

The company has continued in business for many decades, offering a variety of vinyl play dolls as well as limited editions of dolls for collectors. It is, however, the quality of the hard plastic dolls of the 1950s, along with the fine composition dolls of earlier years, that makes Effanbee's reputation secure as one of America's finest doll manufacturers.

Effanbee Doll Corporation 49

15" Effanbee Patsy with cloth body, circa 1927. She has sleep eyes, open mouth with teeth, mohair wig, composition shoulder, head and full arms and legs, with a cloth body. The shoulder plate is marked "EFFANBEE/PATSY" in a half circle. She wears a wool coat and hat with embroidery trim that may be original ($400+). *Doll from the collection of Marge Meisinger. Photograph by Carol Stover.*

Advertisement for EffanBee Durable Dolls from December 1931, which appeared in a magazine called *Junior Home Companion*. The new Lamkin (spelling different than later ads) baby doll was pictured along with a Patsy doll. The advertisement also offered a coupon to be used by little girls to enroll their Patsy doll in the new Patsy Doll Club. Members would receive a Certificate of Membership, Membership Button, and the new club magazine. Two cents in postage was the only cost to join.

15" Effanbee ©Lambkins baby doll, circa 1931. The doll has sleep eyes, open mouth, molded hair, composition head, nearly full arms and legs, and a cloth body. She is marked on the neck "© Lambkins." The doll has unusual curved arms and legs ($250+). *Doll and photograph from the collection of Ellen Cahill.*

6" Effanbee composition Wee Patsy Fairy Princess set from 1935. This boxed Wee Patsy was made to tie in with the publicity generated by the national tour of the Colleen Moore Castle dollhouse. Moore was one of the most popular film stars of the silent screen. Her dollhouse took nine years to complete. After the tour, the Castle was given a permanent home at the Museum of Science and Industry in Chicago, Illinois. The original set also included panties for the checked dress and perhaps an outfit to be sewed. The inside cardboard is missing as is the doll's original pin. The box is marked "Copyright, 1935, by Fleischaker & Baum New York." Another similar set which featured Wee Patsy was a sewing set that included clothing to sew (boxed set $500+).

Effanbee Patsy advertisement carried in the Montgomery Ward Toys Spring and Summer catalog in 1931. She was 13-1/2" tall and sold for $2.65. The copy said that Ward's was the only place Patsy was sold by mail. *From the collection of Marge Meisinger.*

6" Effanbee composition Wee Patsy dolls, circa early 1930s. The dolls have painted eyes, closed mouths, molded hair and shoes and socks, and are jointed at shoulders and hips only. They are marked "Effanbee/Wee Patsy." The dolls are dressed in their original boy and girl outfits. These dolls were first introduced in 1932. They also came in a dollhouse series which included a maid and a "black" cook ($200 each with wear).

7" composition Effanbee Baby Tinyettes, circa early 1930s. These were first introduced in 1933. They have painted eyes, closed mouths, molded hair, and curved baby legs. They are marked on their backs "Effanbee/Baby Tinyette." The dolls are wearing what appears to be their original clothing, including diapers, cotton kimonos, and booties. The Baby Tinyettes were sold individually, in pairs, or in sets of three or five dolls. The five doll set tied in to the Dionne Quints, although they were not authorized dolls. In advertisements for the dolls, they were called "Patsy Tinyettes" and the original boxes and tags carried that name. The dolls are pictured in an unidentified swing and high chair (dolls $500-600 pair, furniture $75-100). *From the collection of Jan Hershey.*

8" Effanbee composition rare Patsy Tinyette doctor and nurse dolls, circa mid 1930s. The dolls were made with the same head and body mold as the Tinyette Babies but the legs on these dolls are straight. They have painted eyes, closed mouths, molded hair, and are fully jointed. They are marked "Effanbee" on their heads and "Baby Tinyette" on their bodies. They are all original except they are missing the printing from their caps that used to read "Patsy Doctor" and "Patsy Nurse" (not enough examples to determine a price). *From the collection of Jan Hershey.*

8" composition Effanbee Patsy Babyette, circa late 1930s-1940. She has sleep eyes, closed mouth, molded hair, and is fully jointed. She is marked "Effanbee" on her head and "Effanbee/Patsy Babyette" on her body. This doll was originally introduced in 1933. She is all original, complete with her Effanbee heart shaped bracelet. This doll was also made with a lambskin wig ($350). *From the collection of Jan Hershey.*

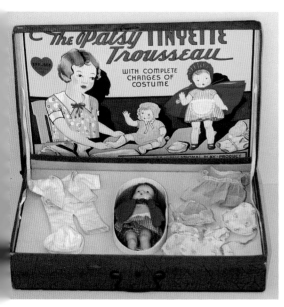

8" Effanbee Patsy Tinyette packaged in a red imitation leather cardboard suitcase, circa mid 1930s. The doll was made using the Baby Tinyette head and body with added toddler straight legs. The lid of the suitcase reads "The Patsy Tinyette Trousseau WITH COMPLETE CHANGES OF COSTUME." In a heart motif is printed "An/EFFanBEE/Play Product."

The packaged Patsy Tinyette set consisted of the composition doll plus four outfits of clothing. These included a Red Riding Hood costume, a sailor suit and hat, and two organdy dresses with matching bonnets. The one piece bloomer underwear that matched the Red Riding Hood dress could also be used as a playsuit. Underwear for the two dresses may be missing from the set. The doll was marked "Effanbee" on the back of her head and "Effanbee/Baby Tinyette" on the back of her body (set $600+).

52 Effanbee Doll Corporation

9" composition Effanbee Patsyette, circa late 1930s-1940. She has painted eyes, closed mouth, molded hair, and is fully jointed. She is marked "EFFanBEE/Patsyette Doll" on the back of her body. She is all original wearing a tagged dress. She is pictured with an Effanbee Patsyette trunk filled with Patsyette clothing (doll $375, empty trunk $150+). *Doll and trunk from the collection of Jan Hershey.*

8-1/2" composition Effanbee pair of Patsy Babyette dolls, circa late 1930s-1940. The dolls have sleep eyes, closed mouths, molded hair, and are fully jointed. They are marked "Effanbee" on the heads and "Effanbee/Patsy Babyette" on their backs. They are wearing their original clothing but their shoes and socks have been replaced. The boy may have originally had a hat ($175-200 each in this condition).

Effanbee "Tousle-Tots" were advertised in the N. Shure Co. catalog in 1941-42. The dolls were actually Patsy Babyette dolls with added lambskin wigs. The dolls could be purchased individually or in pairs. *From the collection of Marge Meisinger.*

9" composition Effanbee Patsyette, circa 1935. This Patsyette has a braided mohair wig. She also has her original Effanbee heart bracelet which isn't shown in the picture. She was probably dressed as Anne from "Anne of Green Gables" and was originally wearing a hat ($250 with some wear).

Effanbee Doll Corporation 53

9" Patsyette twins by Effanbee, circa 1931. The Patsyette dolls were introduced in 1931. They are original except for their missing hats ($650 pair). *From the collection of Jan Hershey.*

11" composition Effanbee Patsy Jr (also known as Patsykins), first introduced in 1930. She has painted eyes, closed mouth, molded hair, and is fully jointed. She is marked on the back "EFFANBEE/PATSY JR/DOLL." She is wearing her original tagged dress, matching one piece underwear, black leatherette shoes, and an Effanbee metal heart bracelet. Her dress tag features a gold heart with red lettering inside reading "EFFanBEE/DURABLE/DOLLS" and under the heart is printed "MADE IN U.S.A." ($375). *From the collection of Jan Hershey.*

Matching boy and girl outfits for the Effanbee Patsyette dolls ($50 each). *From the collection of Jan Hershey.*

Right:
14" composition Effanbee Patsy first introduced in 1928. She has painted eyes, closed mouth, molded hair, and is fully jointed. She is marked on her back "EFFanBEE/PATSY/Pat. Pending." She is in near mint condition and wears what appears to be her original untagged green and white organdy dress with white organdy collar and her original shoes ($400-500 in near mint condition). *From the collection of Jan Hershey.*

Left:
12" composition Effanbee Patsy Baby dolls. Both dolls have sleep eyes, closed mouths, molded hair, curved baby legs, and are fully jointed. The doll on the left is marked "Effanbee/Patsy Baby" on the back of her head and "Effanbee/Patsy Baby" on her back. She wears her original tagged clothing. The tag reads "Effanbee Durable Dolls/Made in U.S.A." Her shoes have been replaced. She was marketed as Patsy Babykin. The doll on the right has a replaced body. The head may have originally come on a rubber body which collapsed. She is wearing an original cotton pique dress. The Patsy Baby dolls were introduced in 1932. Most of the early dolls had blonde hair (dark haired doll with some wear $250). *From the collection of Jan Hershey.*

Effanbee Doll Corporation

Original Patsy clothing. The navy double breasted coat is tagged "Glad Togs/Seattle." The Beach Pajamas are tagged "Niko." This firm usually made Patsy clothes for the trunk sets. The 1946 reissued Patsy was dressed in the Pinafore (beach pajamas $35-40, others $50 each). *From the collection of Jan Hershey.*

14-1/2" composition Effanbee Patsy-Patricia, circa late 1930s. She has sleep eyes, closed mouth, molded hair, and is fully jointed. The back of her neck is marked "EFFANBEE Patsy." Her back is marked "EFFANBEE/Patricia." The doll is all original including her metal heart bracelet. She has never been played with and her costume is crisp and like new. The Patricia doll was first introduced in 1932 ($400+ with some paint flecking on hand).

14" composition Effanbee Patsy, circa 1933-34. She has sleep eyes, closed mouth, mohair wig, and is fully jointed. She wears her original, dress, bonnet, and underwear and retains her Effanbee bracelet. Wigs were added to some of the Patsy Family dolls by 1933 and sleep eyes had been included for the largest dolls in 1932 ($400+) *Doll from the collection of Marge Meisinger. Photograph by Carol Stover.*

14-1/2" composition Effanbee Patricia doll with "Magic" hands, circa 1940. She has painted eyes, closed mouth, molded hair, and is fully jointed. She is marked "EFFANBEE Patsy" on her head and "EFFENBEE/Patricia" on her back. She has magnets in her hands so she can hold different objects. Several other Effanbee dolls also had this feature (see text). This doll has been redressed in an early Patsy family dress ($250+ redressed).

15" composition Effanbee Patricia "Little Eva," circa late 1930s. She has sleep eyes, closed mouth, human hair wig, and is fully jointed. The marked "Patricia" doll was costumed to represent the "Little Eva" character from the Harriet Beecher Stowe book, *Uncle Tom's Cabin* (a Little Eva doll in mint condition with her tag was sold at auction for $1,400). *From the collection of Jan Hershey.*

John Plain featured an Effanbee Patsy Joan which contained a music box in its 1940 catalog. The doll had sleep eyes, closed mouth, molded hair, a composition head, shoulder plate, full arms and legs, and a cloth body that contained the music box. *From the collection of Marge Meisinger.*

16" composition Effanbee Mary Lee-Patsy Joan, circa 1935. This Patsy Joan variant has sleep eyes, open mouth with teeth, a mohair wig in braids, and is fully jointed. She is marked "Mary Lee" on the back of her neck and "EFFANBEE/Patsy Joan" on her back. She was marketed circa 1934-35 to tie in with the *Anne of Green Gables* film. She originally had a hat but otherwise is all original including her Effanbee heart bracelet. Her skirt is removable to reveal her underwear playsuit. Patsy Joan was first introduced in 1931 ($300 with some wear to costume and missing hat).

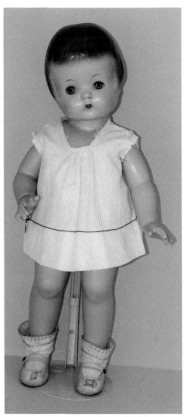

19" composition Effanbee Patsy Ann. She has sleep eyes, closed mouth, molded hair, and is fully jointed. She is marked on the back "EFFANBEE/PATSY ANN © Pat. #1283558." She is wearing her original tagged dress and one piece underwear. The tag reads "EFFANBEE/DOLL/FINEST & BEST" in an oval and underneath "MADE IN USA." Her shoes and socks and Effanbee bracelet are also original. The bracelet reads "F & B Durable Dolls." The Patsy Ann was first introduced in 1929 ($450+). *From the collection of Jan Hershey.*

56 Effanbee Doll Corporation

Effanbee Patsy Lou paper doll from the Whitman #976 set published in 1939. Barbara Lou, another Effanbee doll, was also made in paper doll form for the set (see Dewees Cochran chapter). The original Effanbee Patsy Lou doll was introduced circa 1931 and was 22" tall (cut whole set of paper dolls $125+).

29" Effanbee Patsy Mae, the largest of the Patsy family dolls. These dolls were introduced in 1934 and always came with cloth bodies. She has sleep eyes, closed mouth, human hair wig, composition swivel head, shoulder plate, arms and legs, and cloth body. She is marked on the head "Patsy Mae" and on her body "Effanbee Lovums © Pat No. 123558." (undressed $700). *From the collection of Jan Hershey.*

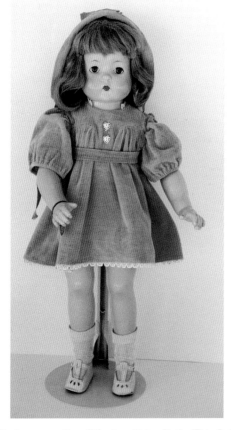

27" all composition Effanbee Patsy Ruth. This Patsy family member was introduced in 1934 with a cloth body. The all composition doll was first made circa 1935. She has sleep eyes, closed mouth, real hair wig, and is fully jointed. She is marked "EFFANBEE/ Patsy Ruth" on her head. This doll is circa 1939. She is all original wearing her rose velveteen dress and matching hat. Her shoes are like the ones worn by the Patsy Lou paper doll from 1939 ($1,400+). *From the collection of Jan Hershey.*

Some of the boxes for the Patsy Ruth dolls were marked "Patricia Ruth". This box gives the information that these dolls had human hair wigs and repeats the company slogans "The Doll With The Golden Heart" and "An EffanBee Durable Doll." (box $100+).

Right:
Patsy Ann Her Happy Times by Mona Reed King. Photographs by G. Allan King, published by Rand McNally & Co., Chicago, 1935. This is the 1937 edition. The story tells about Patsy Ann's daily activities and pictures the doll in various outfits ($40-45).

Booklet called *Aunt Patsy Tells a Story*, issued by the Patsy Doll Club, 45 Greene St., New York. Copyright by Fleischaker and Baum. It is a fiction story about the making of the Dy-Dee Baby ($35).

Right:
The Montgomery Ward Christmas catalog from 1940 also featured the Dy-Dee dolls, along with a separate layette for $1.89. The copy said that the set was made by Effanbee and came in different sizes to fit 11", 15" and 20" dolls. The most expensive set was priced at $3.45. Included in the layette were a dress, bonnet, slip, pajamas, wrapper, bathrobe, booties, undershirt, diapers, rubber panties, blanket, crib pad, feeding spoon, bottle, safety pins, wash cloth, and a cardboard case. *From the collection of Marge Meisinger.*

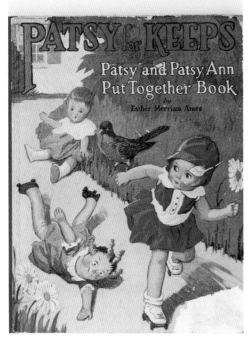

Patsy for Keeps, Patsy and Patsy Ann Put Together Book by Esther Merriam Ames. Illustrated by Arnold Lorne Hicks, published by Sam. Gabriel Sons & Company, New York City, © 1932. The child was to illustrate the book by cutting out pictures from the included gummed sheets and placing the pieces into the proper places in the book ($45-50).

The John Plain catalog featured the Effanbee Dy-Dee doll in its catalog for 1937. The dolls came in sizes of 11", 13", 15", and 20". They had hard rubber heads, soft rubber bodies, sleep eyes, molded hair, and were "drink and wet" dolls. The dolls could be purchased with several different sizes of layettes. *From the collection of Marge Meisinger.*

58 Effanbee Doll Corporation

13" Effanbee Dy-Dee doll, first introduced in 1934. She has sleep eyes, open mouth, molded hair and ears, hard rubber head, soft rubber body, and is a "drink and wet" doll. She is marked on her back "EFFANBEE/DY-DEE BABY/PAT.NO 1859-485/OTHER PAT.PEND." The doll has been redressed. She is pictured with the *DY-DEE DOLL'S DAYS* book. It was written by Peggy Vandegriff with photographs by Lawson Fields and was published by Rand McNally & Co., Chicago, 1937 (doll redressed and worn $50-75, book $35+).

The N. Shure Catalog for 1941-42 pictured the "new" Dy-Dee Baby. Applied soft rubber flexible ears had been added to the design along with open pink lined nostrils. The new model could sip from a spoon and blow bubbles. The doll still retained the hard rubber head and soft rubber body and came in sizes of 11", 15", and 20". *From the collection of Marge Meisinger.*

Dy-Dee's Diary copyright by Fleischaker and Baum, New York. This little book came with various Dy-Dee dolls, circa 1939. It is like a baby book that includes spaces to write about the baby's arrival, guests, playmates, etc. ($25).

The Montgomery Ward catalog for 1942 featured a full page advertisement for the Dy-Dee doll and her accessories. This doll had a lambskin wig as well as a full layette. The dolls and layettes came in sizes of 11" priced at $8.79, 15" for $12.50, and 20" costing $21.00. A separate layette was still available as was a Dy-Dee Bathinette. *From the collection of Betty Nichols.*

Effanbee Doll Corporation 59

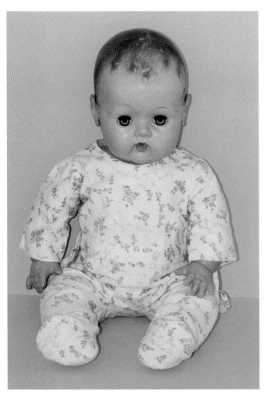

Right:
15" composition "1607 Indian Squaw With Braids" from the 1939 replica series of Effanbee Historical Dolls. She has painted eyes, closed mouth, braided wig, and is fully jointed. The doll is all original and has the brown skin of an Indian. All of these dolls were marked "Anne Shirley" on their backs ($500-600.) *Doll from the collection of Marge Meisinger.* Photograph by Carol Stover.

20" Dy Dee doll, circa 1949. She has a hard plastic head, rubber body, sleep eyes, and open mouth. Her rubber ears are applied. The doll is marked on the back of her head "Effanbee." Her back is marked "EFF-An-BEE/DYDEE BABY/ U.S.PAT.-1857-485." Other patent numbers are also given for several other countries. She is dressed in "Dy-Dee Baby" print flannel one-piece pajamas. The words "Dy-Dee Baby" and flowers decorate the material. The outfit is an Effanbee product but not original to this doll. She can drink and wet and blow bubbles ($150-200).

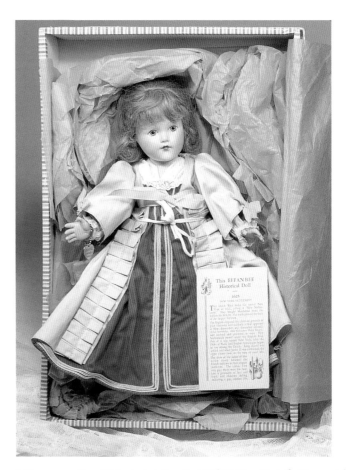

An Effanbee Touslehead doll was featured in the Montgomery Ward catalog for 1939. The doll was 18" tall with a composition head, arms, legs, and a soft body. She had sleep eyes and a soft fur skin wig. The doll had a tiny phonograph in her body which played "Now I Lay Me Down To Sleep." It sold for $12.95. *From the collection of Marge Meisinger.*

15" composition Historical doll called "1625 New York Settlement." She has painted eyes, closed mouth, human hair wig, and is fully jointed. She is all original and still has her tag and box. All of these dolls are marked "Anne Shirley" on their backs (sold at auction for $735). *Courtesy of Frashers' Doll Auctions, Inc.*

Effanbee Doll Corporation

15" Effanbee replica "1685 Later Carolina Settlement" Historical doll. She is all original with her Effanbee bracelet. The larger 20" Effanbee Historical dolls used the closed mouth American Children molds and are much more scarce and expensive ($450-500).

15" replica "1888 Settling the West" Historical doll. She is all original. Although the 20" dolls were dressed in fancy materials including taffeta and satin, the replica dolls were dressed in cotton ($450-500). *Courtesy of JJ's Doll Company.*

15" replica "1840 Covered Wagon Days" Historical doll. She is all original with her Effanbee bracelet. There were thirty dolls in the series. Only the costumes were different ($450-500). *Doll from the collection of Jan Hershey.*

15" replica "Today" doll (representing circa 1939). She is all original including her Effanbee bracelet. The necklace may have been added. Her costume was designed by Chanel. This doll was the last in the series of dolls ($500).

"Sonja on Skates," 16 inches tall. More than just a fine doll, EFFanBEE has captured the very expression of her famous namesake. Her face is that of an older person instead of the ordinary child. Every detail is perfect even to her tiny lacquered finger nails. Sonja is an all composition doll having almost real glass eyes and lashes, and a curled wig of human hair. Beautifully dressed in a white taffeta skating ensemble, maribeau trimmed around edge of skirt; combination underwear to match; white taffeta tam with maribeau pom-pom. Opera length white silk hose; high laced white shoes with skates attached.
No. 36N312. Each..........$7.50

The Effanbee "Sonja on Skates" was advertised in the N. Shure Co. catalog for 1938. This composition doll was also known as the "Ice Queen" and was made to take advantage of the publicity generated by the ice skating star Sonja Henie. Although the doll was not authorized by the famous star, many little girls probably thought they were receiving a real Sonja doll when this doll was marketed. The head of the doll was molded from the Dewees Cochran open mouth doll design. *From the collection of Marge Meisinger.*

The N. Shure Co. advertised an Effanbee "Heartbeat" doll in 1941. The doll was a Lovums model with an added feature that allowed the owner to listen to the "baby's heartbeat" by using a special stethoscope. The doll came in sizes of 17" and 19". *From the collection of Marge Meisinger.*

16" Effanbee Lovums baby dolls, circa 1943-45. The dolls have sleep eyes, open mouths, lambskin wigs, composition heads, shoulder plates, arms and legs, and cloth bodies. The dolls are marked on the back of the shoulder plates "Effanbee Lovums ©." The dolls belonged to sisters so the doll on the left has a blue outfit and blue eyes, and the doll on the right has a pink dress and bonnet and brown eyes. The dolls are missing their shoes and socks, otherwise they are all original. When the skirts are removed, the dolls are dressed in rompers ($250+ each). *Childhood dolls from the collections of Bonnie McCullough and Marilyn Pittman. Photograph by Marilyn Pittman.*

Right:
Montgomery Ward featured the Effanbee Little Lady doll in its Christmas catalog in 1940. The composition doll, called "Magic Hands," had magnets in her hands so that she could hold different objects. She was 15" tall and was wearing a pink organdy print party dress with a pink organdy undie, extra long white rayon stockings, and pink leatherette slippers. From the collection of Marge Meisinger.

Several Effanbee composition Little Lady dolls were featured in the Montgomery Ward Christmas catalog for 1942. These dolls were the Anne Shirley dolls under a different name. All of the dolls had sleep eyes, closed mouths, human hair wigs, and were fully jointed. Doll A. came in a 20-1/2" size dressed in a coat and hat, B. was 21-1/2" tall and was dressed in a robe, bra, and panties, C. came in an 18" size and was dressed in a dancing dress, D. came in three sizes of 18", 21-1/2", and 28" and was costumed in a long gown of pink marquisette with a black bodice, E. was 20-1/2" tall and her gown was embroidered net with a pink underskirt. The 28" model was the largest Little Lady made. *From the collection of Betty Nichols.*

18" composition Effanbee Little Lady as pictured in the Montgomery Ward Christmas catalog for 1942. She has sleep eyes, closed mouth, real hair wig, and is fully jointed. The doll is marked on the neck "Effanbee USA." She is all original, dressed in a South American dancing dress made of organdy with a taffeta lining and green sandals (original with bracelet $300+). *From the collection of Marge Meisinger. Photograph by Carol Stover.*

27" Effanbee composition Little Lady. This is the largest size of this doll. She has sleep eyes, closed mouth, wig, and is fully jointed. The doll is marked on her head "EFFANBEE." She is all original including her bracelet. She wears a formal, underwear, long white stockings, and sandals ($400+). *From the collection of Jan Hershey.*

"Little Lady Jigsaw Puzzle" © 1943 by Fleischaker & Baum, New York. The printing on the box reads "An/Effanbee/Play product/America's Doll Family." The same picture was used for a two page color advertisement in the Montgomery Ward 1942 Christmas catalog. The Little Lady dolls in this picture ranged in size from 15"-21-1/2" tall ($40-45 missing one piece).

Effanbee Doll Corporation 63

20" Effanbee Glamour doll called Patricia. She has a composition head (mounted on wooden neck piece), composition hands, and cloth body and limbs. She has painted eyes, closed mouth, and—because of the shortages caused by World War II—her hair is made of heavy thread. She is wearing her original costume except that the blouse is missing its anchor decoration and the sash. Her shoes are replaced. The doll has knee joints made in the cloth legs. During these years, doll manufacturers used as much cloth as possible to save on scarce materials ($200).

18" composition Effanbee Little Lady. She has sleep eyes, closed mouth, human hair wig, and is fully jointed. She is marked "Effanbee USA" on the back. She is all original including her long gown, slip, panties, socks and gold sandals. She is similar to several other Little Lady dolls pictured in the 1942 Montgomery Ward Christmas catalog, which included one bride doll and four dolls in long dresses ($300).

Two pages of Effanbee dolls advertised in the Montgomery Ward Christmas Catalog of 1943. Included were three baby dolls using the "Sweetie Pie" head in sizes of 17", 19", and 24", Skippy dolls dressed in soldier and sailor uniforms, and Little Lady dolls dressed as a majorette and a variety of little girl costumes. They ranged in size from 18" to 21" tall.

18" composition Effanbee Little Lady, circa 1943-45. She has painted eyes, closed mouth, yarn hair, and is fully jointed. She is marked on her head "Effanbee" and marked on her body "Effanbee USA." She is all original, wearing her pink formal trimmed in blue, underwear, socks and sandals. The painted eyes replaced the sleep eyes in earlier dolls because of the scarcity of the materials ($300).

Left:
The Montgomery Ward Christmas catalog from 1945 featured several composition Little Lady dolls, including a bride in an 18" size and a doll dressed in an evening gown complete with a "fur" jacket in 18", 21", and 27" sizes. Both of these dolls had the yarn type hair even though the war had ended in August of 1945. The dolls ranged in price from $12.95 for the bride to $39.95 for the 27" Little Lady.

14" composition Effanbee Little Lady. She has sleep eyes, closed mouth, mohair wig, and is fully jointed. Although she is wearing a wedding dress similar to those pictured in the 1945 catalog, she must date from an earlier period as her back is marked "Effanbee/Anne-Shirley." Perhaps she was old stock that was used when parts were scarce during World War II. She is all original including her rayon dress, veil, slip, panties, socks, sandal shoes, and bouquet ($200+).

The Montgomery Ward Christmas catalog was still featuring the Little Lady Effanbee dolls in 1948. The composition dolls came in only two sizes, 15" and 18", and were priced from $7.79 to $11.65. A new "Honey" doll was also pictured in the upper left hand corner. This doll came in a 21" size and sold for $17.95. She had flirty eyes and was an all composition doll. Later, the same mold would be used for Effanbee's popular hard plastic Honey dolls. *From the collection of Betty Nichols.*

8" composition Effanbee "Butin-Nose," circa late 1930s-early 1940s. He has painted eyes, closed mouth, painted hair, and is fully jointed. He is all original wearing a black cotton suit and hat, black leatherette shoes, and white socks (sold at auction with one finger broken for $445). *Courtesy of Frashers' Doll Auctions, Inc.*

The Montgomery Ward Christmas catalog for 1940 pictured an Effanbee Suzette doll that was priced at only 95 cents. The 11-1/4" doll had painted eyes, closed mouth, molded hair and was fully jointed. Her costume was green and white. *From the collection of Marge Meisinger.*

9" composition Effanbee Little Girl doll, circa late 1930-early 1940s. She has painted eyes, closed mouth, molded hair, and is fully jointed. The doll is marked on the back "EFFANBEE/Made in USA." She is all original wearing a dotted swiss and organdy dress, white lace trimmed slip and attached bloomers, original leatherette shoes, and socks ($175+). *Courtesy of JJ's Doll Company.*

11-1/2" composition Effanbee George and Martha Washington, circa 1930s. They were made using marked Suzette dolls. They have painted eyes, closed mouths, "powdered" wigs and are fully jointed. They are all original wearing costumes made to represent the first president and his wife. Many dolls were sold in similar outfits in 1932 to tie in with the Bicentennial celebration of George Washington's birth, but these appear to be newer (sold at auction for $655). *Courtesy of Frashers' Doll Auctions, Inc.*

66 Effanbee Doll Corporation

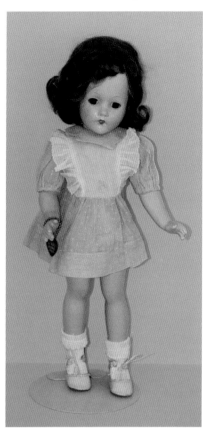

Left:
14" composition Effanbee Suzanne, circa 1940. She has sleep eyes, closed mouth, mohair wig, and is fully jointed. She is marked "Suzanne/Effanbee/Made in U.S.A." She is all original wearing a pink dress and her original "Effanbee Durable Dolls" metal heart bracelet ($250-300).

Right:
12" composition Effanbee portrait dolls called Dancers or Spanish Dancers by collectors. They have sleep eyes, closed mouths, mohair wigs, and are fully jointed. They are all original and are unmarked (pair $450-$500).

12" composition Effanbee Portrait Bride and Groom, circa 1942. The dolls are all original in their box. They have sleep eyes, closed mouths, mohair wigs, and are dressed in bridal attire. The dolls are unmarked (pair with box $600+). *Dolls from the collection of Marge Meisinger. Photograph by Carol Stover.*

12" composition Effanbee unidentified portrait doll. She is all original and probably represents a character from a nursery rhyme ($225).

Four 12" composition Effanbee portrait dolls. They include a Ballerina, the female Dancer, Gibson Girl (missing her hat), and Little Bo-Peep (sold at auction for $200-235 each with light craze). *Courtesy of Frashers' Doll Auctions, Inc.*

Effanbee Doll Corporation

The Montgomery Ward 1943 Christmas catalog devoted a full page in color to the Effanbee Brother and Sister set of dolls. The pair of dolls sold for $6.95.

The Effanbee Brother and Sister dolls featured in the 1943 catalog had painted eyes, closed mouths, yarn hair, composition heads and hands, and cloth bodies, legs, and arms. These dolls were made during World War II and the cloth parts and yarn hair were used because of the scarcity of materials. The boy is 16" tall and the girl is 12" high. Both dolls are marked "Effanbee" on their heads. The dolls are all original (pair $500-600). *Childhood dolls of the author received for Christmas in 1943.*

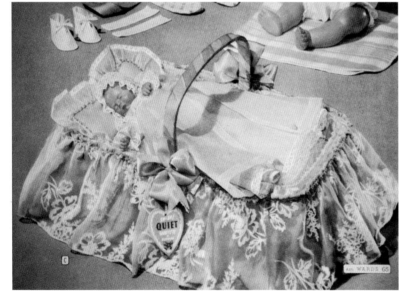

The Montgomery Ward Christmas catalog for 1945 advertised an Effanbee sleeping Babyette, complete with its own basket for $16.95. The doll was 12" tall and her eyes were painted asleep.

12" Effanbee Sleeping Babyette doll as pictured in the 1945 catalog. She has a composition head and hands, with cloth body, arms, and legs. The doll has molded hair and painted features including eyes painted asleep. She is all original, dressed in a rayon marquisette christening dress, coat, and bonnet. Alexander, Ideal, and Madame Hendren all marketed dolls of this type during the 1940s ($250-300). *From the collection of Phyllis Young.*

68 Effanbee Doll Corporation

Effanbee's "Sweetie Pie" doll was featured in the Montgomery Ward 1945 Christmas catalog. The dolls came in sizes of 19" for $11.50 and 24" for $15.95. They had flirty sleep eyes, closed mouths, lambskin wigs, composition heads, arms, legs, and cloth bodies with cry boxes. The dolls were dressed in rayon marquisette dresses and matching bonnets.

10" Effanbee baby doll, circa late 1940s. She has sleep eyes, closed mouth, lambskin wig, composition head, and one piece cotton stuffed latex body. She is marked "EFFANBEE" on her head. She is all original except for her replaced bonnet. She wears a diaper, socks with pink ribbon ties, slip, dress, and bonnet. Her body has darkened some with age ($125+).

The Children's Activities magazine carried an Effanbee doll advertisement in its November 1947 issue. Pictured were the Dy-Dee baby, Little Lady Bride and Bridesmaid, Sweetie Pie, Mickey, and the Candy Kid Sister and Brother dolls. A coupon was provided to order an Effanbee catalog.

Left:
16" Effanbee Bright Eyes, circa mid 1940s. He has flirty eyes, closed mouth, lambskin wig, composition flange head, full arms, curved legs, and a cloth body and upper legs. He is marked on the back "© 194? EFFANBEE." He is all original. This same head was also marketed as "Sweetie Pie," "Tommy Tucker," and "Mickey."

14" Effanbee Mickey dolls, circa 1947. The dolls have flirty sleep eyes, closed mouths, and molded hair. They have composition flange heads and hands and cloth bodies and limbs. They are marked on their heads "EFFANBEE/Made in USA." The original paper tag reads "I Am/Mickey/ With Moving Eyes/An/Effanbee/ Durable Doll."The back of the tag says "ANew/Playmate/For.../ From.../May You And Mickey/Have Many Happy Times Together/ Made in/USA." The dolls are all original ($300+ each). *From the collection of Jan Hershey.*

20" Effanbee doll using the same head. The author received this doll for Christmas in 1944 and always called it "Sweetie Pie" as a child. Collectors now call him Mickey. The doll was a war-time product and it was made with a composition head and hands and a cloth body and limbs. He has flirty sleep eyes, closed mouth, and a mohair wig. He is marked "EFFANBEE" on his head. He is all original ($350+).

13" Effanbee composition Candy Kid Brother and Sister dolls, circa 1946-47. They have sleep eyes, closed mouths, molded hair, and are fully jointed. The dolls are marked "EFFANBEE" on the heads and bodies. The tag reads "I AM AN/EFFANBEE/DURABLE DOLL/ THE DOLL/WITH/SATIN SMOOTH/SKIN." The dolls are all original (pair $700-800). *From the collection of Jan Hershey.*

Eugenia Doll Company

The New York based Eugenia Doll Company issued quality dolls during the 1940s and 1950s. The firm marketed baby, toddler, little girl, and young women dolls. The earlier models were made of composition while the later dolls were made of hard plastic. Many of the dolls were featured in the Montgomery Ward Christmas catalogs during the late 1940s. In 1948, Montgomery Ward pictured five different Eugenia dolls called Personality Pla-Mates. Two of the dolls were hard plastic while the other dolls were composition.

In the 1946 Montgomery Ward Christmas catalog, the "Eugenia Miniatures" were advertised. These composition dolls ranged in size from 5-1/2" to 6-1/2" tall and sold for $1.59 to 2.49 each. A 7" baby was also pictured, selling for $3.29. These dolls were apparently produced to compete with the popular Nancy Ann products.

6-1/2" composition Eugenia Saturday doll with painted features, a mohair wig, and movable arms. She has painted shoes and still has her original box and wrist tag. The tag reads "Eugenia Doll Co." on one side and "A Touch of Paris" on the other. The box is labeled, "A Touch of Paris Copyright 1945, Eugenia Doll/ Saturday/Our Sweet Eugenia is so Gay,/ Because Today is Party Day." ($25-35).

The Montgomery Ward Christmas Catalog for 1946 featured a full page ad for small "Eugenia Miniatures" composition dolls. The dolls came in sizes of 5-1/2" for $1.59 and 6-1/2" for 2.49. A 7" baby was also offered for $3.29.

6-1/2" composition Eugenia Sunday doll, circa mid 1940s. She is exactly like the Saturday doll and the dolls pictured in the Montgomery Ward catalog except for her costume ($25-35).

Eugenia Doll Company

A 17" Eugenia composition little girl doll advertised by Montgomery Ward in their Christmas catalog for 1947. The fully jointed doll had sleep eyes, a closed mouth, and a human hair wig. She was dressed in a pink rayon dress and sold for $9.50.

This set of Eugenia brother and Sister Twins was featured in the Montgomery Ward catalog in 1948. The dolls were 12" tall with sleep eyes, closed mouths, and hair made from a heavy thread. The heads and hands were composition while the rest of the dolls were made of cloth. The pair sold for $7.75. The boy doll had dark hair and was dressed in blue. The little girl had yellow hair and was dressed in pink. The dolls are very similar to the Effanbee Brother and Sister dolls from 1943. *From the collection of Betty Nichols.*

18" composition Eugenia doll, circa 1946-47. She has sleep eyes, a closed mouth, a mohair wig, and is fully jointed. She is wearing her original tagged clothing. The tag reads "Eugenia Doll Co./America's Finest Dolls/Made in U.S.A." ($150).

12" Eugenia Sister doll as pictured in the Montgomery Ward catalog in 1948. She has sleep eyes, closed mouth, thread hair, composition head and hands, and a cloth body, legs, and arms. She is all original and near mint. She once had a brother doll with dark hair, dressed in blue ($150).

Eugenia Doll Company

12" composition Eugenia baby doll, circa 1947. She has sleep eyes, closed mouth, molded hair, and is fully jointed. She has a toddler composition body and this same doll was used as the model for a set of twins as well as a little boy. The doll is all original with her original box. The dress tag reads "Eugenia Doll Co./Made in U.S.A./America's Finest Dolls." The box is labeled "Eugenia dolls/Adorable Dolls/A touch of Paris Copyright 1945. Eugenia Doll Co. New York-Paris." ($175-200).

Advertisement in the Montgomery Ward 1949 Christmas catalog for a Eugenia Sandra Pla-Mate. The caption describes the doll as being all plastic with sleep eyes and a mohair wig. This same doll was also produced in composition and was pictured in the Montgomery Ward catalog in 1948 made of this material and dressed in a different costume. *From the collection of Betty Nichols.*

Georgene Novelties, Inc.— Madame Hendren

Georgene Hendren was born in Denver, Colorado in 1876. She married James P. Averill in 1914 and the couple would work in the doll business, under various company names, for many decades. Georgene had begun making dolls in 1913. Her first dolls were composition character dolls dressed in felt. In 1915, Georgene began using the trade name Madame Hendren and founded the Averill Manufacturing Co. in New York. Her husband and brother both helped in the company. In the early years, she produced dolls with composition heads, composition hands, and stuffed bodies. She also designed bisque heads that were made by Alt, Beck & Gottschalck and Gebruder Heubach. Bonnie Babe, a baby doll, was Georgene's most famous bisque design. Dolly Record was also a very famous product from the 1920s. This composition and cloth doll contained a record player in its body.

The firm continued to market dolls into the 1930s under various names including Madame Georgene, Inc. and Georgene Novelties. As the decade began, the company was still selling its popular all composition Dimmie and Jimmie dolls. These dolls had a body twist construction so they could turn or bend at the waist. The firm was also advertising both little girl and baby models of the traditional "Mama" dolls with composition heads and limbs and cloth bodies.

One of the unusual dolls produced by the company in the early 1930s was a breathing doll. When pressure was applied, air came out of the composition doll's nose and mouth.

The firm followed the trend in the early 1930s when they issued a "look alike" Patsy doll called Peaches. This composition doll had molded hair and came with either painted or sleep eyes. The dolls were issued in sizes of 14" or 17". A larger 19" version included a cloth body.

Several famous artists created dolls for Madame Hendren. Included were Grace Corey, Grace Drayton, Maud Tousey Fangel, and Harriet Flanders. The Corey and Drayton designs were produced in the 1920s, while the cloth dolls designed by Maud Tousey Fangel were sold by Georgene Novelties in the late 1930s. These dolls sold very cheaply in 1938 (under $1.00). The original lithographed faces were designed and painted by Fangel. Most of the dolls were made of printed cloth and the clothes on most of them were very simple—a gathered skirt and a bonnet. Several different designs were sold including a baby made of sateen. These came in 8" and 11" sizes and were named "Sweets." Larger little girl dolls were made in 14" and 16" sizes and were labeled Peggy Ann. A Snooks doll was also produced. Very nice reproductions of some of the dolls have been made in recent years.

The Harriet Flanders dolls were issued circa 1937. These all composition dolls came in sizes of 12" and 17". The original tag read "Little/Cherub/A/Georgene Doll" and the dolls were marked "Harriet © Flanders 1937." A book called *Little Cherub*, with illustrations by Harriet Flanders, came with each doll. The dolls came in various costumes including a skier, ice skater, Indian, and a toddler.

Another interesting Madame Hendren composition doll was Baby Yawn, circa 1946. This doll had a composition head and hands on a cloth body. It was an unusual doll because the eyes were painted closed and the mouth was molded in a permanent yawn.

By the early 1940s, Georgene firms seemed to be concentrating most of their effort on their famous cloth dolls, sold under the Georgene Novelties banner. The company had been making "International" cloth dolls for many years. These dolls were issued in native costumes from countries around the world and were often made in pairs.

Georgene Novelties became the exclusive manufacturers of Johnny Gruelle's Raggedy Ann and Andy dolls in 1938. The dolls continued to be made by that firm until 1963. The first of the Georgene dolls had the red noses outlined in black with black shoe button eyes. The earliest of these dolls had a separate label for each doll identifying it as a Raggedy Ann or a Raggedy Andy doll. Later, only one label was used which identified the dolls as "Raggedy Ann and Andy Dolls." (see Character Dolls chapter).

Georgene Novelties also produced a series of cloth comic dolls in their later years. Although many references say the dolls were made in 1944, it appears they may not have been made until the 1950s. Although a Georgene Little Lulu doll's original purse is marked with a "Marge 1944" copyright, the doll was purchased by the author in 1955. A similar Alvin doll from the same comic is tagged "Copr. 1951 Marjorie H. Buell." A matching Nancy and Sluggo set of dolls, from the same period, were also produced. These dolls were based on the "Nancy" comic strip by Ernest Bushmiller (see Character Dolls chapter).

Georgene Averill retired to California where she died in Santa Monica in 1961. She left an unusual doll legacy for collectors because her dolls included those made of bisque, composition, celluloid, and cloth. The current prices for her dolls range from very reasonable (cloth International) to quite expensive (bisque Bonnie Babe). Every collector should be able to find an Averill doll they can afford and enjoy.

Right:
The 1931 Montgomery Ward Spring and Summer catalog advertised Madame Hendren "Mama" dolls in a variety of sizes. The dolls had composition heads, arms, and legs, and cloth bodies. They featured real hair wigs, sleep eyes, and lashes. The dolls came in sizes of 18", 20", 22", and 24" at prices from $4.98 to $9.45. *From the collection of Marge Meisinger.*

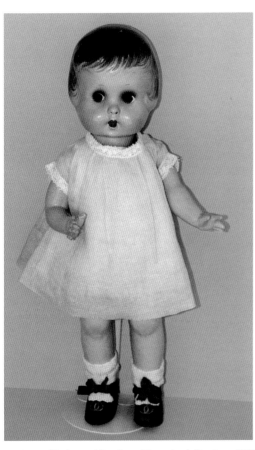

14" composition Madame Hendren Dimmie doll, circa 1930. She has painted eyes (with lots of eye shadow), a closed mouth, molded hair, and is all composition. She has a special swivel construction that allows her to twist or bend at the waist. She is jointed at the neck, shoulder, breast, waist, and hips. A similar Jimmie boy doll was also marketed. Her dress appears to be original but her bonnet is missing ($350+). *Courtesy of JJ's Doll Company.*

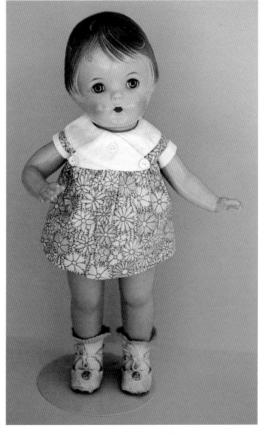

14" composition Madame Hendren Peaches. She has blue tin sleep eyes, a closed mouth, and molded hair. There are two molded curls on the top of her head. She is fully jointed and appears to be all original. She is marked "A CO INC" on her head (Averill Co., Inc.) This doll was produced in the early to mid 1930s to compete with the Effanbee Patsy doll ($250+). *From the collection of Jan Hershey.*

Georgene Novelties, Inc.— Madame Hendren 75

12" unmarked all composition little girl doll with her original clothes and trunk, circa 1930s. She has painted eyes, a closed mouth, and molded hair. One of the dresses that came with the trunk carries a Madame Hendren tag so she may be a Hendren product; not all of their dolls were marked. Besides the Hendren dress, the doll's wardrobe includes bonnet, two dresses (one of which appears to be very similar to the dress on the Dimmie doll), pique coat and hat, pajamas, underwear, and roller skates ($200+).

20" Madame Hendren brown baby doll, circa 1930s. She has brown sleep eyes, open mouth with two teeth, and molded black hair. The body and upper legs are cloth while her flange head is composition, as are her full arms and lower legs. The body contains a crier. She is marked "Baby Hendren" on the back of her neck. She has been redressed ($150-200).

Madame Hendren tagged dress and underwear from the trunk that came with the unmarked composition doll. The tag is sewn to the outside of the back of the dress. It reads "Madame Hendren/DOLLS/Everybody Loves Them." The outfit is made very cheaply with no fasteners and the hem turned under with raw edges left showing.

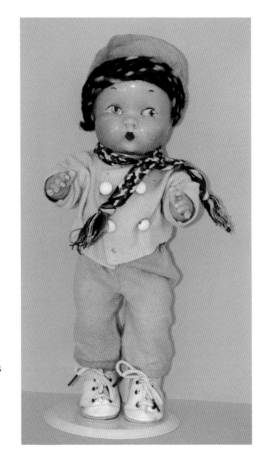

12" composition Little Cherub skier, circa 1937. He has painted eyes, closed mouth, molded hair, and is fully jointed. He is marked "Harriet © Flanders 1937." The dolls were created by Harriet Flanders but came with a tag identifying them as Georgene products. A companion ice skating little girl doll was also made. This doll wears his original clothing with replaced shoes. It is likely that he originally had skis ($150-175).

Georgene Novelties, Inc.— Madame Hendren

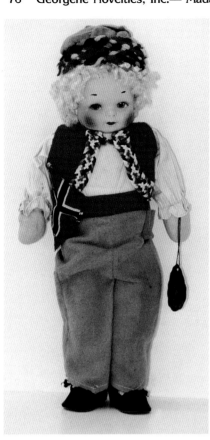

Left:
14" Georgene Novelties cloth Norway doll from the International series, circa 1930s. He still has his original tag and box. He has a buckram type face (coarse cotton cloth stiffened with glue) with painted features, applied eyelashes, and yarn hair. He is all original, complete with his Norway flag. Many other companies also made this type of cloth doll. The Georgene dolls have mitten shaped hands, real feet with shoes (scallop design), and were not stitched up the back by machine. Original tag reads "A PRODUCT OF/GEORGENE NOVELTIES, Inc./NEW YORK, N.Y./MADE IN U.S.A." ($100-125).

10" cloth Topsy & Eva Georgene Novelty doll. Two doll heads (one light and one dark) share one body. The long reversible skirt covers the head not in use. The doll has flat painted faces and yarn and painted hair. Each head has its own set of arms with mitten hands. The original tag reads "ART IN CLOTH DOLLS/Topsy and Eva/AN ORIGINAL GEORGENE NOVELTY." ($150).

18" pair of Johnny Gruelle's Raggedy Ann and Andy dolls made by Georgene Novelties. The Georgene firm produced these dolls from 1938 until 1963. The earliest dolls' noses were outlined in black and they had shoe button eyes. These dolls date from the early years with the tag reading "Johnny Gruelle's own 'Raggedy Ann' Doll/ Georgene Novelties, Inc., New York City." The dolls are original except that Ann may be missing her apron (pair $1,200+). *Dolls from "Suzie's Museum of Childhood" at Bluebird Farm in Carrollton, Ohio. Photograph by Marilyn Pittman.*

13" cloth Georgene Novelty Brownie Scout, circa late 1940s. She has a buckram type face with painted eyes, closed mouth, and embroidery thread hair. She is hand stitched up the back and wears shoes with scalloped edges. This doll has been seen with an original Georgene tag. She is all original, wearing her Brownie uniform, including her beanie ($75+).

E.I. Horsman Dolls, Inc.

E.I. Horsman Dolls, Inc. was founded by Edward Iseman Horsman, circa 1865, in New York City. The firm dealt in children's toys, games, tricycles, skates, and sleds.

Although the Horsman Company had imported doll heads and bodies earlier, around 1900 the company was also marketing the popular Campbell Kid and Baby Bumps dolls of composition.

The company continued its success in the 1920s with its Baby Dimple dolls. Louis Amberg & Son sold Horsman their composition doll line in 1930. Some of that firm's dolls were marketed by Horsman using other names. One of these was the 14" all composition little girl doll with molded hair and a twist waist that Horsman called Peggy. A 1930 ad listed sizes as 12", 14", and 20". The firm also marketed the Amberg 10" toddler body twist dolls under the Horsman name.

Other dolls issued by the firm in the early 1930s included several sizes of Patsy look-alikes with molded hair and sleep eyes in four sizes. The Montgomery Ward Catalog from 1931 pictured the four dolls, called Babs (12"), Sue (14"), Jan (17"), and Nan (20"). The dolls were priced from 98 cents to $3.98 each. Other dolls advertised in this two page spread were the Baby Dimples dolls in three sizes and the Rosebud little girl "Mama" doll, also in three sizes. Other baby dolls were also pictured, ranging in size from 13" to 26".

During the 1930s, the Horsman firm developed a line of dolls they called Buttercup. Although most of these dolls had composition heads, arms, and legs, the name was also used for a series of "drink and wet" dolls. The head on these dolls was hard rubber and the body was soft rubber. They had sleep eyes and molded hair.

In 1937, Horsman produced some of their most innovative composition dolls when the company brought out a line of "Whatsit" all composition dolls. Naughty Sue had molded hair, painted eyes, and was 14" tall. Roberta was also 14" tall, with molded hair and sleep eyes. Other composition dolls from this same period included the 12" Jo-Jo, which came with sleep eyes and either molded hair or a wig. Jeanie, another unusual model, had a composition head, arms and legs, a cloth body, molded hair, and sleep eyes. This doll is similar to the sister doll that was made as part of a brother and sister pair of dolls also from 1937. These larger dolls also had cloth bodies and were marketed in 21" (Brother) and 24" (Sister) sizes.

The Horsman firm continued its innovative doll designs in 1938 when the company marketed its Sweetheart doll line. These large dolls were probably made to compete against the Deanna Durbin dolls. They came in three sizes: 20", 24", and 28". The N. Shure Co. catalog for 1938 priced the dolls from $5.80 to $9.80 each. They were made of composition, except for the arms, which were made of hard rubber. The clothing was designed to make the dolls appear to be teen-agers. The dolls had sleep eyes, mohair wigs, and open mouths. According to information on the original hang tag, the dolls were sculptured by Ernesto Peruggi.

In 1940, the E.I. Horsman firm was purchased by Regal Dolls and the name was changed to Horsman Dolls, Inc. After the sale of the company, the new owners seemed content to produce ordinary baby, toddler, and little girl dolls. The firm's dolls in 1941 included the Rosebud Mama Dolls, a line called Baby Precious (which featured large baby dolls from 18" to 24" tall), and all composition little girl dolls labeled with the Bright Star name. The Bright Star line of dolls came in 19" and 21" sizes and featured sleep eyes and mohair wigs. The Spiegel catalog from 1941 featured Rosebud as well as an all composition little girl called "Pigtail Sally" and a skating doll which appeared to be the same doll dressed in a different costume. The company continued to sell these types of dolls during most the decade of the 1940s. Increasingly, the firm marketed its dolls to the cheaper market. The Montgomery Ward Christmas catalog for 1949 featured a line of Horsman baby dolls that looked very much like the dolls from 1941 but the arms and legs were made of "Plastic Baby Skin" instead of composition. The babies ranged in size from 15" to 25" and featured sleep eyes and either mohair or molded hair. They were sold at prices from $2.95 to $7.95.

Horsman did have one more composition success when they received authorization to produce the Campbell Soup Kids dolls in 1948 (see Character Dolls chapter).

By 1950, Horsman had begun making some of its little girl line in hard plastic and baby dolls of latex but it wasn't until 1952 that Horsman completely dropped their composition dolls from its catalogs. The dolls from Horsman's earlier years will remain as the most desirable examples of fine composition dolls from this firm.

Horsman advertisement in the *Child Life* magazine for December 1930 pictures three of the dolls offered by the company. The Dolly Rosebud and Baby Dimples lines had proved to be very successful. The Peggy doll was one of the composition dolls that Horsman acquired when they bought out the Louis Amberg & Son line in 1930. The Amberg doll had been called Sue.

E.I. Horsman Dolls, Inc.

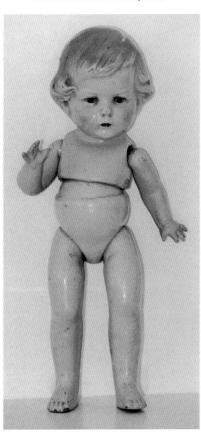

Left:
14" composition body twist jointed doll sold by both Louis Amberg & Son (Sue) and Horsman (Peggy). She has painted eyes, closed mouth, and molded hair. She has an unusual joint in her middle that allows her to twist or bend. She is marked "AMBERG/PAT. PEND./LA&S. © 1928. The doll was marketed by Horsman after they purchased the composition doll line from Louis Amberg & Son in 1930 ($300+ redressed).

Right:
10" body twist doll which came in both black and white versions, circa 1930. She has painted eyes, closed mouth, and molded hair. She is jointed at the shoulders and lower waist. Marked "Pat. APPLD" on her back. She still has her original one-piece cotton print playsuit. She was probably first developed by Amberg but continued to be sold by Horsman after they purchased the composition line of dolls from Amberg ($175+ with original outfit).

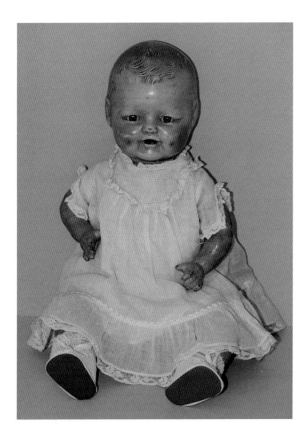

16" Horsman Baby Dimples. This example has a composition flange head, full composition arms, and curved composition lower legs. The body and upper legs are cloth. She has metal sleep eyes, open mouth with teeth, and molded hair. Marked ©/E.I.H. CO. INC. on the back of her head. She wears her original tagged dress, slip, and attached panties but is missing her bonnet. Her shoes and socks have been replaced. Dress tag reads "HORSMAN/DOLL/M'F' D'IN U.S.A." ($150-185 for doll with some cracks).

15" Horsman Buttercup, circa 1932. This version of the Buttercup doll was featured in the Sears catalog of 1932. She has a composition head, full composition arms and legs, and a cloth body. The doll has sleep eyes, molded hair, a closed mouth, and has been redressed ($150-175 redressed). *Doll and photograph from the collection of Ellen Cahill.*

E.I. Horsman Dolls, Inc. 79

Advertisement in the Montgomery Ward Christmas catalog for 1933 featuring Horsman dolls. This line of dolls was produced to compete with the Effanbee Patsy. Included were the 12" Babs, 14" Sue, 17" Jane, and 20" Nan. The dolls were priced from 98 cents to $3.98. All were made of composition, with sleep eyes, and molded hair. *From the collection of Marge Meisinger.*

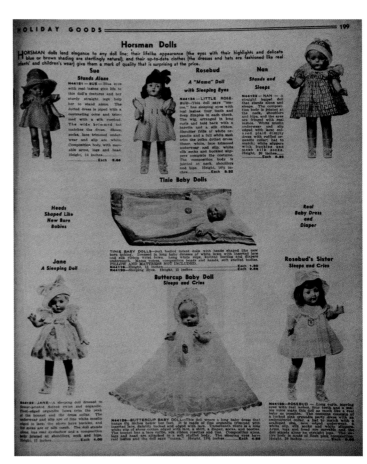

McClurgs 1934 Fall and Winter catalog also advertised Horsman dolls. Included were two of their popular baby dolls called Tinie Baby and Buttercup. Both dolls had unique composition heads. The Tinie Baby was supposed to represent a new baby while the Buttercup Baby doll had a fuller, rounded face. The Tinie Baby in this ad came in 13" and 15" sizes. The Buttercup Baby was 13-1/2" tall. *From the collection of Marge Meisinger.*

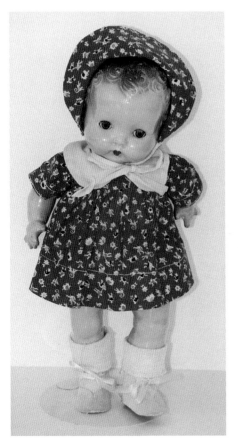

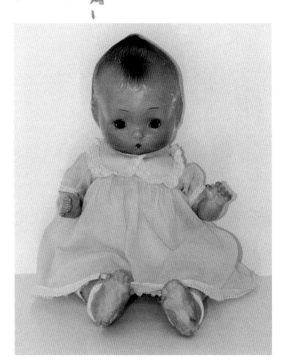

Left:
12" Horsman Jo-Jo all composition doll. She has tin sleep eyes, a closed mouth, and molded hair. Marked on the back of her neck "JO-JO/© 1937 Horsman." The fully jointed doll wears her original clothing. The shoes and socks have been replaced ($175-200).

Right:
15" Horsman Jeanie, circa late 1930s. She has sleep eyes, closed mouth, and molded hair with a peak on top. Her flange head, full arms, and lower swinging legs are composition while her body and upper legs are cloth. She has a crier box. She is marked on her back, "©/Jeannie/Horsman." Her clothing appears to be original but the shoes and socks have been replaced. Jeanie looks very much like the much larger Sister doll marketed during the same time period ($200+).

80 E.I. Horsman Dolls, Inc.

Horsman Brother (21") and Sister (23") dolls from 1937. Both dolls have sleep eyes, closed mouths, and molded hair. They have composition flange heads, arms, and lower legs. Their bodies and upper legs are cloth. They are marked on their necks, "Brother/1937" and "Sister/1937." Both dolls have been redressed and had some re-painting done ($350-450 for set in this condition). *From the collection of Phyllis Young.*

On the left, a 16" composition Horsman Art Doll. She has sleep eyes, open mouth, mohair wig, and a fully jointed body. She is all original wearing a pink taffeta dress, attached panty, and bonnet. Her tag reads "A/Genuine/Horsman/Art Doll." ($275-300). On the right, a 16" composition Roberta by Horsman. She has sleep eyes, closed mouth, molded hair in coiled side braids, and is fully jointed. She is marked "Roberta © 1937 Horsman." She has been redressed ($300+). *Courtesy of Frashers' Doll Auctions, Inc.*

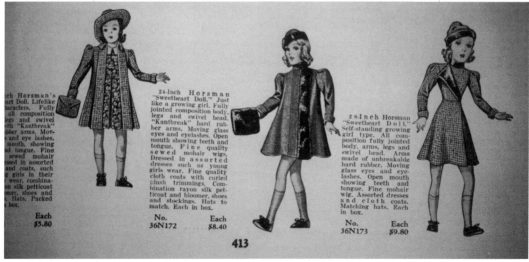

N. Shure Co. 1938 advertisement for the Horsman Sweetheart Dolls. The dolls had sleep eyes, open mouths, mohair wigs, and were made of composition except for their hard rubber arms. They came in three sizes: 20", 24", and 28". They were supposed to represent teen-age girls and were dressed accordingly. They may have been made to compete with Ideal's Deanna Durbin dolls. *From the collection of Marge Meisinger.*

E.I. Horsman Dolls, Inc. 81

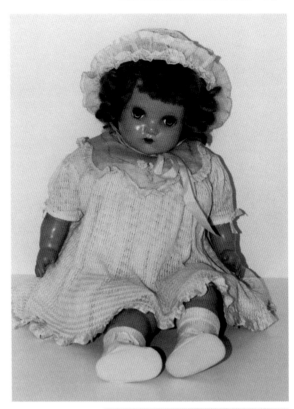

Right:
23" Horsman baby doll, circa 1940s. She has sleep eyes with eye shadow, open mouth with teeth, mohair wig, a composition head, lower arms, and legs, and a cloth body (with crier box) and upper limbs. She is marked "Horsman/Doll" on the back of her head. She is wearing factory made clothing which may be original. She has replaced shoes and socks ($125-150).

12" baby doll, circa mid 1940s, thought to have been made by Horsman. Although the doll is unmarked, its original owner remembers it being a Horsman product. She has sleep eyes with eye shadow, closed mouth, molded hair, and a composition head, arms, and legs, with cloth body. She is wearing her original dress and bonnet but is missing her shoes and socks ($125-150). *Childhood doll of Bonnie McCullough. Photograph by Marilyn Pittman.*

21" Horsman baby doll, circa 1940s. She has sleep eyes with eye shadow, closed mouth, red mohair wig, composition head, lower arms and legs, cloth body (with crier box) and upper limbs. She is marked "Horsman Dolls" on the back of her head and is wearing her original clothing except for shoes and socks ($150+).

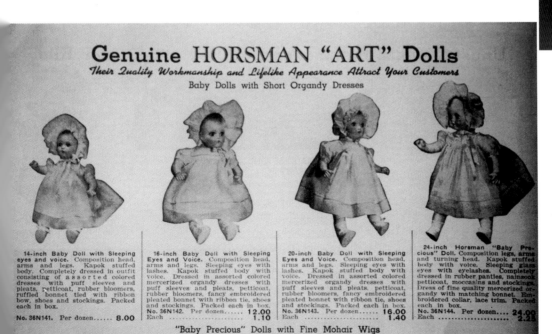

Advertisement from the N. Shure Co. catalog for 1941 featuring Horsman baby dolls. The dolls came with molded hair or wigs in sizes of 14", 16", 20", and 24". The trade name for the dolls was "Baby Precious." *From the collection of Marge Meisinger.*

82 E.I. Horsman Dolls, Inc.

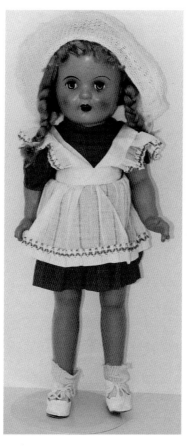

20" composition unmarked doll, circa early 1940s. Although she is not marked, she appears to be one of the dolls produced by Horsman. She has lots of eye shadow which the firm used in the 1940s. She has sleep eyes, open mouth with teeth, mohair wig in pig tails, and is fully jointed. She is all original unless her hat is a replacement ($175).

Horsman dolls were also advertised by the Chicago Wholesale Co. in their 1946-47 Fall and Winter catalog. Both baby dolls and little girl dolls were pictured with and without wigs in sizes up to 21" tall. *From the collection of Marge Meisinger.*

The Spiegel catalog for 1941 advertised a Horsman skating doll called Jessie. She came in 15" or 18" sizes. The composition doll was fully jointed with sleep eyes and a real hair wig braided into pigtails. The dolls were priced at $2.39 and $2.98. *From the collection of Marge Meisinger.*

19" Horsman baby doll, circa 1940s. She looks very much like the doll in the upper right corner in the Chicago Wholesale Co. ad. She has sleep eyes, closed mouth, mohair wig, composition flange head, lower arms and legs, cloth body (with a crier box) and upper limbs. She is marked "Horsman/Doll ©" on the back of her neck. She is all original ($200+).

E.I. Horsman Dolls, Inc. 83

14" composition Bright Star, circa late 1940s. She has sleep eyes with eye shadow, open mouth with teeth, and a mohair wig. The fully jointed doll is not marked. She is wearing her original wedding gown and veil, which is similar to one pictured on a hard plastic doll in the Horsman 1950 catalog. The lace on the skirt has been added to cover some worn places in the costume ($100-125).

Horsman used the Bright Star trade name for many years. Pictured is an advertisement from the Chicago N. Shure Co. catalog from 1941 featuring Bright Star dolls. These dolls came in 19" and 21" sizes with sleep eyes, open mouths, and mohair wigs. They were dressed in long dresses, which was unusual for Horsman. *From the collection of Marge Meisinger.*

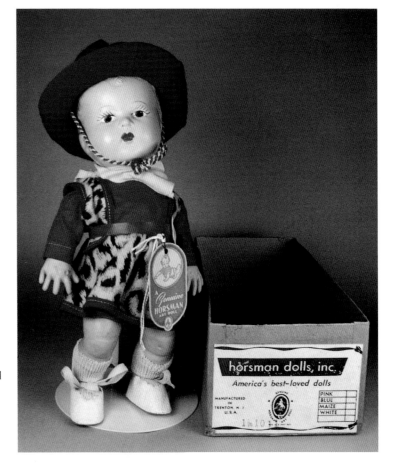

10" Horsman composition-latex doll, circa 1949-50. She has painted eyes, closed mouth, and molded hair. Her head is composition and she has a natural stuffed rubber body. She is all original with her box and tag. Her tag reads "A/ Genuine/ HORSMAN/ART DOLL." She is dressed in her original cowgirl outfit that was later used on a 13" doll, with sleep eyes, which was issued as one of a set of twins ($200+). *Doll and photograph from the collection of Carol J. Stover.*

84 E.I. Horsman Dolls, Inc.

12" Horsman Campbell Kids, circa 1948. The matching dolls have painted eyes, closed mouths, molded hair, and are fully jointed. Their shoes and socks are painted and they are unmarked. The dolls were based on the advertising characters used to sell Campbell Soup. The boy probably wore a cook's cap originally and although the girl wears a factory made sunsuit, it is not known if it is original ($400-500 for pair in this condition).

22" brown Horsman baby doll with "plastic skin," circa 1949. She has sleep eyes with eye shadow, open mouth with teeth, black mohair wig, and a composition head. Her lower arms and legs are of stuffed vinyl and she has a cloth body with a crier. She is marked "Horsman Doll" on the back of her neck. She is wearing her original nylon dress and bonnet but her shoes and socks have been replaced ($175-200).

Advertisement from the Montgomery Ward 1949 Christmas catalog which features Horsman dolls with "Soft Plastic Baby Skin." The copy says "the arms and legs are made of soft Vinylite plastic." The heads were composition with sleep eyes and some models had mohair wigs while others had molded, painted hair. The dolls came in sizes from 14"-25" tall and ranged in price from $2.95 to $7.75. *From the collection of Betty Nichols.*

Ideal Toy Company

Ideal Toy Co. had its beginning when Morris Michtom and his wife began producing stuffed bears above their candy store in Brooklyn, New York, in 1903. The couple called the bears "Teddy Bears" after President Theodore Roosevelt. Michtom went into partnership with Aaron Cone in 1907 and the Ideal Novelty Co. was born. In 1912, this partnership ended and Michtom changed the name of the company to the Ideal Novelty and Toy Co. The firm produced many innovative dolls in the teens and '20s including Liberty Boy in 1918 and a two-faced Soozie doll in 1923.

The Ideal Flossie Flirt little girl doll appeared on the scene in 1924 and was such a hit, it was still being marketed into the 1930s. The doll came in various sizes (from 14" to 28") and models over the years. Her most advertised characteristic was her flirty eyes, which moved from side to side.

The company also produced a baby doll called Tickletoes with these same type eyes. This doll was sold throughout the 1930s in various sizes including 16", 18", 20", 22" and 24". Most of the dolls were made with a composition head, cloth body, and rubber arms and legs. The rubber on most of the dolls deteriorated over time so not many excellent examples remain.

Other popular Ideal babies of the 1930s included Ducky, Cuddles, and Snoozie. Ducky was another doll with a composition head and rubber arms and legs but he also had a rubber body. This doll was a "drink and wet" model. Ducky was introduced in 1932 and was sold until 1939. He came in sizes of 11", 13", 15", and 17".

Cuddles was another trade name used for Ideal baby dolls of the 1930s. The older version, being sold in the early 1930s, was made with a composition head, and legs, a cloth body, an unusual rubber shoulder plate, and rubber arms. The later dolls had composition heads and limbs, with cloth bodies. All the dolls had sleep eyes and open mouths with teeth. They were made in 14", 16", 20", 22", 25", and 27" sizes.

The Ideal Snoozie baby dolls were also popular in the early 1930s, from 1933-35. This baby doll had a composition head with hands and legs made of rubber. Her body was cloth. She had holes in her nostrils and the advertising stated that when the doll was hugged "you can feel her soft breath upon your cheeks." She came in sizes of 14", 16", and 24".

Throughout the 1930s, Ideal continued to market little girl dolls as well as babies. One of the most popular was Miss Ginger. She had a composition head and body and some models featured new "all-direction jointed arms and legs." This model was advertised in the John Plain catalog for 1937 and came in sizes of 20", 23", and 26".

Of course, the Shirley Temple dolls, marketed by Ideal from 1934 to 1939, were the best selling dolls of the decade. Given the success of these dolls, Ideal produced other celebrity dolls, including Judy Garland, Deanna Durbin, Mortimer Snerd, and Baby Snooks, which also sold well (see Personality Dolls chapter). During these same years the firm marketed dolls based on Walt Disney films, including *Snow White and the Seven Dwarfs* and *Pinocchio* (see Character dolls chapter).

Even with the popularity of Ideal's personality and character dolls, the company continued to produce outstanding baby and girl models. One of the most lasting was the Betsy Wetsy doll introduced in 1937-38. The earliest of these dolls came in sizes of 11", 13", 15", 17", and 19". They featured hard rubber heads and soft rubber bodies and limbs. The dolls were "drink and wet" models. They could be purchased alone or packaged in a suitcase with various accessories. The Betsy Wetsy dolls, in many sizes and models, were marketed by Ideal off and on until the 1980s. The dolls' material changed from rubber to vinyl but the basic premise of a "drink and wet" doll never varied.

Other popular Ideal baby dolls of the era included "Baby Beautiful" and "Princess Beatrix." These dolls were sold from 1938 until the early 1940s. Baby Beautiful was made in 16", 18", 20", and 27" sizes. The doll had a composition head and limbs, cloth body, and flirty sleep eyes. Princess Beatrix came in sizes of 14", 16", 22", and 26", and also had a composition head and limbs with cloth body (see Personality Dolls chapter for more information).

When the Madame Alexander Doll Co. began producing their Sonja Henie skating dolls, other large doll firms competed by marketing skating dolls of their own. Many of these dolls were purchased by unknowing adults thinking they were Sonja dolls. Ideal's model was called "Queen of the Ice." She came in sizes of 13", 16", 18", and 20". She was an all composition, fully jointed doll with sleep eyes and a mohair wig. The dolls were sold from the late 1930s until the early 1940s.

Ideal also continued to make many other all composition little girl dolls during the 1930s and 1940s. Many of these were advertised in mail order catalogs but the maker was not identified. Betty Jane was an example of this type of doll. In 1941, the N. Shure Co. advertised these dolls in 15", 18", 20", and 22" sizes. Vogue Dolls, Inc. also dressed some of these same dolls and marketed them under their name.

Coquette was a rather more unusual all composition Ideal doll from the late 1930s. She was made in a 9" size with painted eyes and a mohair wig.

Composition bride dolls were popular with all the major doll companies in the 1940s and Ideal marketed several different models. Many of the Ideal brides had chubby figures compared with the trim Alexander examples. In the 1945 Montgomery Ward Christmas catalog, a whole Ideal wedding party was featured, with a 21" bride, a 22-1/2" bridesmaid, and a 14" flower girl.

During the World War II years, doll production was slowed due to the shortage of materials. Ideal did market several interesting military dolls (see Military Dolls chapter) and quite a number of interesting cloth dolls (see Cloth Dolls chapter).

With the end of the war in 1945, Ideal's doll manufacturing returned to full production. Several new baby dolls were marketed to the public at this time. Included were the Plassie dolls with hard plastic heads and shoulder plates, cloth bodies, and composition arms and legs. The advertising states that the dolls' heads would tilt or turn. The 1945 Sears Christmas catalog sold these dolls in sizes of 17", 19", and 22".

The same Sears catalog also featured two unusual Ideal baby dolls that came as a pair for $9.98. They were called Sleepy-Time Twins in the advertising but are sometimes referred to as Yawn and Dawn. One baby was modeled in a sleeping mode while the other doll was produced as a yawning infant. The dolls were 15" tall.

Although the "Magic Skin" latex had been used by Ideal earlier in the 1940s, its was not until after the war that the product gained acceptance. Quite a number of the Ideal baby and toddler dolls from 1946-50 had bodies made of this new material. One line of dolls was called "Magic Skin Dolls." They were advertised in 14", 16", and 18" sizes by Sears in 1946. Other dolls featuring these bodies included the Baby Coos family (Baby, Brother, and Sister) and Sparkle Plenty. The Coos dolls made a cooing sound when the body was pressed. The problem with many of these "Magic Skin" bodies was that with the passing of time, most of the early ones lost their flesh color and turned dark brown.

As the decade of the 1940s neared its end, the Ideal company ceased making composition dolls and began production of the nearly indestructible hard plastic and vinyl dolls.

Beginning in 1947, the Ideal firm began scoring hit after hit with their new lines of dolls. This success continued throughout the decade of the 1950s, and Ideal was perhaps the most successful company in the doll industry during this time period. Some of their hits included Betsy Wetsy, Toni, Saucy Walker, Shirley Temple, Miss Revlon, and Patty Play Pal.

The Ideal company was run by family members through the early 1980s. According to Judith Izen in her book *Collectors Guide to Ideal Dolls*, the firm was sold to the Columbia Broadcasting System in 1983 for $58 million. The line then merged with CBS's Gabriel Toy Co. and was called Ideal-Gabriel. Viewmaster then bought the Ideal trade mark but after a short time they sold out to Tyco Industries, Inc. in 1989.

The toys made by the original Ideal company are still remembered fondly by today's collectors and many of the company's composition and hard plastic dolls remain as some of the finest dolls ever made.

An advertisement in the Sears Fall and Winter catalog from 1930-31 featured Flossie Flirt dolls made of composition and cloth with rubber arms. The dolls' eyes could flirt as well as wink. *From the collection of Marge Meisinger.*

Left:
19" Ideal Flossie Flirt crier doll, circa 1930. She has a composition head, arms and lower legs, cloth body and upper legs, tin flirty sleep eyes, open/closed mouth, mohair wig, and a crier box. She is marked "IDEAL" in a diamond. She appears to be all original ($200+). *Doll and Photograph from the collection of Ellen Cahill.*

The Ideal Snoozie doll was advertised by the Blackwell Wielandy Co. in their Holiday Catalog in 1935-36. The baby dolls were offered in three sizes of 13-1/2", 16", and 24". *From the collection of Marge Meisinger.*

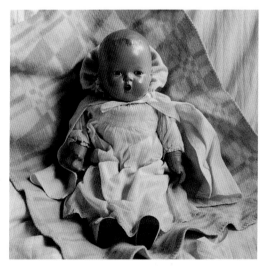

15" Composition Ideal Snoozie baby doll. The doll has tin sleep eyes, open/closed mouth with teeth, pierced nostrils, molded hair, composition head and lower legs, rubber lower arms, cloth body with crier, and cloth upper arms and legs. Marked "B. Lipfert" on neck. The doll has been redressed ($200-250). *Doll and photograph from the collection of Ellen Cahill.*

Ideal 87

Left:
Ideal's Miss Ginger advertised in the John Plain catalog for 1937. She came in sizes of 20", 23", and 26". The composition doll had flirty sleep eyes and a real hair wig. Her description also mentioned the "new all-direction jointed arms and legs." *From the collection of Marge Meisinger.*

Right:
12" Ideal Fanny Brice (Baby Snooks) Flexy doll. She has molded hair, painted features, composition head and hands, wooden body and shoes, and flexible wire limbs. The doll also has a molded hair loop with a ribbon. The doll is all original wearing a two piece costume and retains her tag, which reads "FLEXY/AN IDEAL DOLL/Fanny Brice's/BABY SNOOKS" ($300+ in this condition). *Doll and photograph from the collection of Judith Armitstead, the Dollworks.*

Ideal Betsy Wetsy "Drink and Wet" doll as pictured in "The Toy Parade" in 1938. The doll was sold in sizes of 10-1/2", 11", 13", and 15", and was priced from $2.95 to $5.95. She came in her own suitcase which included the doll, an "organdie dress and bonnet, slip, bootees, diapers, and bottle." *From the collection of Marge Meisinger.*

11" Ideal Betsy Wetsy, circa 1938-39. She has a hard rubber head, soft rubber body, tin sleep eyes, open mouth, and molded hair. She is marked "Ideal Doll" on her head and body. She could "drink" her bottle and "wet" her diaper through a hole in her buttocks. She is pictured in her original suitcase along with several of her original handmade outfits. Her rubber body has "melted" in places, as happened to most of these dolls ($50-75 in this condition, including suitcase).

88 Ideal

Miss Tickletoes, another Ideal baby doll, was featured in the John Plain catalog for 1940. Various versions of this doll were on the market for many years. The doll came in sizes of 15-1/2", 18", and 22". She had a composition head, rubber arms and legs, and a cloth body. When the leg that had a voice was pressed, the doll cried. She had flirty eyes, open mouth with teeth, and molded hair. *From the collection of Marge Meisinger.*

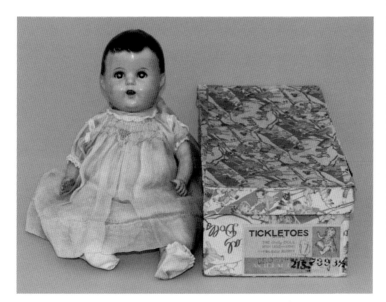

16" Ideal Tickletoes, circa 1940. She has a composition head, sleep eyes, open mouth with teeth, molded hair, curved rubber arms and legs, and a soft cloth body with crier. When she was new, the doll cried when the leg that had a voice was pressed. She is unmarked. She wears her original organdy dress, slip, and shoes and socks. Her rubber pants have fallen to pieces and she is missing her bonnet. She is pictured with her original box. The printing on the box reads "TICKLETOES/THE ONLY DOLL WITH LEGS AND ARMS/of TRU-FLESH RUBBER/AN IDEAL DOLL" ($125 with some rubber "melted").

The Sears catalog for 1939 advertised the Ideal Ducky Baby for a price of $1.79. This 11" all rubber doll had sleep eyes and was a cheaper Ideal "drink and wet" doll. She came with a suitcase, cradle, and a 17 piece layette for only $1.79. *From the collection of Marge Meisinger.*

Right:
Two of Ideal's baby dolls were advertised in the John Plain catalog in 1940. Both Baby Beautiful and Princess Beatrix were made with composition heads, arms and legs and flirty sleep eyes and molded hair. Baby Beautiful came in sizes of 16", 18" and 20" and Princess Beatrix was sold in 16" and 22". *From the collection of Marge Meisinger.*

Ideal also produced a composition ice skating doll to compete with Alexander's Sonja Henie dolls. It was advertised in the N. Shure Co. catalog for 1938. The doll came in two sizes, 13" and 18", and was called "QUEEN OF THE ICE." *From the collection of Marge Meisinger.*

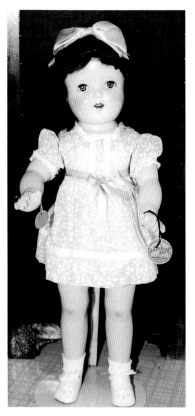

Ideal 22" composition little girl doll. She has sleep eyes, open mouth with teeth, mohair wig, and is fully jointed. This doll is mint with her original box and tag. The tag reads "An/Ideal/Doll/ MADE IN U.S.A. BY/IDEAL NOVELTY & TOY CO./LONG ISLAND CITY/NEW YORK." Her body is marked "Shirley Temple" (not enough examples to determine a price in this condition). *Doll and photograph from the collection of Veronica Jochens.*

Ideal composition Betty Jane dolls were featured in the N. Shure Co. catalog for 1941. The dolls had sleep eyes, open mouths with teeth, and mohair wigs. They came in sizes of 15", 18", 20", and 22". These dolls were also purchased nude by Vogue and costumed and sold under the Vogue name. *From the collection of Marge Meisinger.*

The face of the Ideal little girl, which may be Betty Jane, still retains the original beautiful coloring. To find a composition doll in this mint condition is very unusual. *Doll and photograph from the collection of Veronica Jochens.*

11" Ideal composition Coquette dolls, circa late 1930s-early 1940s. The fully jointed dolls have closed mouths, painted eyes, and mohair wigs. Both are marked "11 Ideal Doll U.S.A." They are wearing their original formals, panties, socks and gold leatherette shoes (Left: $200-225; right: $100 with some wear). *Photograph courtesy of Frashers' Doll Auctions, Inc.*

Montgomery Ward pictured a Bridal Party of Ideal dolls in their 1945 Christmas catalog. The composition dolls included a 21" bride priced at $20.95, 22-1/2" bridesmaids for $17.95 each, and a 14" flower girl for $5.95. The Ideal bride dolls were not as slim figured as the Alexander models.

22" composition Ideal Bride, circa mid 1940s. She has sleep eyes, closed mouth, mohair wig, and is fully jointed. She has her original tag, which reads "AN/IDEAL DOLL/MADE IN U.S.A. BY/IDEAL NOVELTY & TOY CO./LONG ISLAND CITY/NEW YORK." She is all original wearing a satin wedding dress, veil, slip, underpants, shoes, and socks, and carries her bouquet ($300+).

17" Ideal baby doll, circa 1940s. She has a composition flange head, lower arms and legs, cloth body with crier, sleep eyes, open mouth with teeth, and molded hair. She is marked "IDEAL DOLL/ MADE IN U.S.A." on the back of her neck. She is wearing factory made clothing which may be original, replaced shoes and socks ($100-125).

The Ideal Plassie doll was advertised in the Sears Christmas catalog in 1945. The doll featured a hard plastic head, composition arms and legs, and a cloth body. She came in a 17" size priced at $4.98, 19" for $5.98, and 22" for $7.98.

19" Ideal Plassie baby, circa mid 1940s. She has a hard plastic head and shoulder plate, composition full arms and nearly full legs, cloth body and upper legs. The doll has sleep eyes, closed mouth, and molded hair. Her body contains a crier. She is all original and still has her tag. It reads "AN IDEAL DOLL/PLASSIE/PLASTIC HEAD THAT TILTS OR TURNS." The other side reads "AN/ IDEAL DOLL/ MADE IN U.S.A. BY/ IDEAL NOVELTY AND TOY CO./ LONG ISLAND CITY,/ NEW YORK. Marked on the back of her head is "IDEAL DOLL/MADE IN U.S.A./Pat. No 2252077" ($150-175).

Ideal Sleepy-Time Twins pictured in the Sears 1945 catalog. The dolls were 15" tall and had composition heads, hands and legs, and cloth bodies. One doll was pictured asleep while its twin was modeled with its mouth in a yawn. The pair came in a bunting for $9.98.

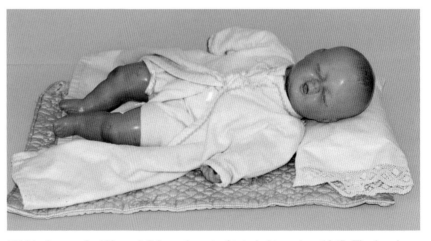

15" Ideal unmarked Yawn doll from the set of twin babies, circa 1945. The head, hands, and legs are composition while the body and upper limbs are cloth. The doll's features are painted with closed eyes and the mouth in a yawn. She is wearing her original diaper and shirt but the robe has been replaced and she is missing her socks ($135-150).

92 Ideal

Ideal's Magic Skin baby dolls were featured in the Sears 1946 Christmas catalog. The dolls came with hard plastic heads, while their bodies and limbs were of synthetic rubber "Magic Skin" stuffed with cotton. They had sleep eyes, closed mouths, and molded hair. In this advertisement, the dolls came in 14", 16", and 18" sizes accompanied by various layettes. The dolls were also issued in a 20" size. *From the collection of Betty Nichols.*

A page of Ideal dolls was advertised in the Chicago Wholesale Co. catalog for Fall and Winter of 1946-47. Included were four sizes of Magic Skin Baby Dolls, five sizes of the Baby Beautiful, three sizes of a baby doll with an applied wig, and 18", 20" and 22" all composition girl dolls. *From the collection of Marge Meisinger.*

14" Ideal Magic Skin baby, circa late 1946-48. She has a hard plastic head, Magic Skin body, sleep eyes, closed mouth, and molded hair. Her arms are jointed but her legs are made in one piece with the body. She is all original complete with her tag, which reads "Magic Skin Baby." Her Magic Skin has darkened with age ($50-75).

21" composition Miss Curity doll thought to have been made by Ideal, circa 1940s. She has sleep eyes, closed mouth, mohair wig, and is fully jointed. She is in mint condition wearing her nurse's uniform, her original underwear, and shoes and socks. She even has original Band-Aids in her pocket. These dolls may have been used in drugstores to promote their Curity products (Mint $600).

Mary Hoyer Doll Manufacturing Co.

Mary Hoyer began her career by designing children's fashions that were to be knitted or crocheted by consumers. Soon the company, located in Reading, Pennsylvania, began selling undressed dolls along with patterns for doll clothing to be made at home. According to Mary Hoyer in her book *Mary Hoyer and Her Dolls*, the first dolls marketed in 1935 were composition body twist models that had been produced by Ideal Novelty and Toy Co. These dolls were to be discontinued by Ideal so Mary Hoyer bought the remaining stock to begin her doll enterprise. When these dolls sold well, Hoyer contacted famous doll designer Bernard Lipfert and hired him to design a doll for the Hoyer Company. The resulting doll was a 14" tall slim girl model with a closed mouth. It was produced by the Fiberoid Doll Co. The doll was sold either undressed or with clothes made by the Mary Hoyer Co.

The earliest of these dolls had painted eyes and the first fourteen hundred dolls were unmarked. Sleep eyes and the familiar trademark circle on the dolls' backs were added later.

The dolls and accessories were marketed through ads in magazines like *McCall Needlework* and were sold through mail order. In 1943, the dolls sold for $2.00 each. The dolls and accessories could also be purchased at the retail shop in Reading, Pennsylvania.

The composition dolls were discontinued in 1946 and new dolls of hard plastic were produced. These new dolls used the same earlier design and were also 14" tall with sleep eyes, a closed mouth, and an applied wig. On the back of each doll was the Mary Hoyer trademark inside a circle.

As styles in dolls changed during the 1950s, the Hoyer company marketed new dolls to meet the competition from other firms. Some of the designs were sold only for a short time while others proved successful and were kept in the Hoyer line for years.

Although dolls were a big part of the Mary Hoyer business, most of the profits came from the sale of clothing and patterns designed for the dolls. The firm sold finished costumes, kits that included materials to make the clothing for the dolls, and booklets called *Mary's Dollies* that included patterns to make outfits for the various dolls. The company also marketed many accessories for their dolls, including trunks, shoes, socks, glasses, skis, skates, wigs, and much more. Mary Hoyer marketed the same patterns for many years, along with new designs which continued to appear on the market. It is difficult to determine when a particular outfit was made, especially if it was sewed, knitted, or crocheted at home.

The Hoyers retired in 1972 but they were not forgotten by the doll world. In 1992, Mary Hoyer was honored with a Lifetime Achievement Award presented by the International Doll Academy and the *Doll Reader* magazine.

Left:
14" composition early Mary Hoyer doll, sculptured by Bernard Lipfert. She has painted eyes, closed mouth, mohair wig, and is unmarked. She is jointed at the neck, shoulders, and hips. She is dressed in an ice skating costume made from a Hoyer pattern ($350-375). *From the collection of Jan Hershey.*

14" composition Mary Hoyer Bride and Groom dolls. These dolls are the later versions featuring sleep eyes. They have closed mouths, mohair wigs, and are fully jointed. They are both marked in a circle on their backs "Original Mary Hoyer Doll." The dolls appear to be original, except for the bride's veil which has been replaced, but the dress is not marked. The Hoyer male dolls are just like the female dolls except for the masculine clothing (Groom $450+, Bride $250-300). *Groom from the collection of Marge Meisinger.*

Mary Hoyer Doll Manufacturing Co.

14" marked composition Hoyer doll dressed in a tagged net and taffeta formal made from a Hoyer pattern. She has an unusual red mohair wig and brown eyes ($450). *From the collection of Jan Hershey.*

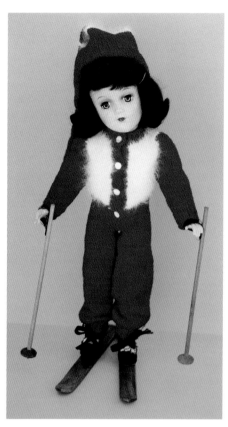

14" marked composition Hoyer doll wearing a Hoyer ski outfit complete with skis and poles ($450). *From the collection of Jan Hershey.*

A Mary Hoyer painted eye composition doll models the Red Cross Nurse costume in an undated Hoyer catalog.

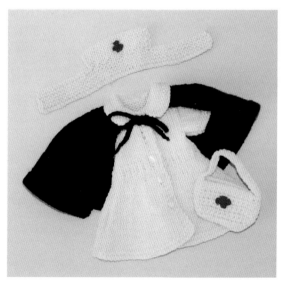

The same nurse outfit is pictured here, featuring the nurse headband, cape, dress, and first aid kit. It was made from a Hoyer pattern ($50-75). *From the collection of Jan Hershey.*

Mary Hoyer Doll Manufacturing Co. 95

Advertisement from *McCall Needlework* magazine for Fall 1950, featuring Mary Hoyer dolls, clothing, and trunks. The dolls were being made of hard plastic by this time. McCall was also offering a pattern to make clothes for the Hoyer dolls. It was #1564 and sold for 35 cents.

Girl and boy costumes made from the Hoyer patterns featured in the *McCall Needlework* magazine for Fall 1950. The kits to make these items sold for 75 cents each ($75-100 vintage set). *Collection of Jan Hershey.*

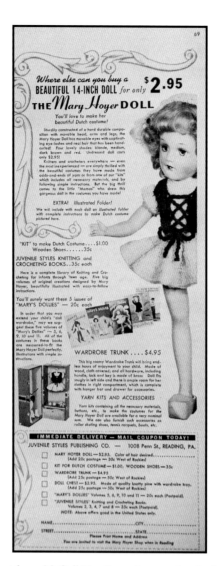

Advertisement from *McCall Needlework* magazine for Summer 1949. It pictured a wardrobe trunk for the Hoyer dolls for $4.95. The undressed hard plastic doll sold for $2.95.

This Mary Hoyer wardrobe trunk is similar to the one pictured in the *McCall Needlework* magazine for Summer 1949. This trunk is filled with Hoyer clothing and accessories. The clothes were made from Mary Hoyer patterns (trunk alone $75-100). *From the collection of Jan Hershey.*

One of several Hoyer designed ballerina costumes is pictured here, along with the shoes. These items probably date from the late 1940s or 1950s when ballerina dolls were so popular ($50-75). *From the Collection of Jan Hershey.*

The aqua and white dress originally came as part of a Dutch outfit that included a hat and wooden shoes. The red and yellow costume was part of a bare back rider outfit. The ice skates and roller skates are also Hoyer products (vintage knitted items $35-50 each, skates $35-50 each). *From the collection of Jan Hershey.*

This negligee came with a matching gown and has been seen with a Hoyer tag. The bra and panties were made from a Hoyer pattern ($75-100 for the three pieces). *From the collection of Jan Hershey.*

The Mary Hoyer firm also offered several varieties of western wear. This set is missing its pants but does include the original hat, boots, guns, holsters, and sweater ($75-100 as pictured). *From the collection of Jan Hershey.*

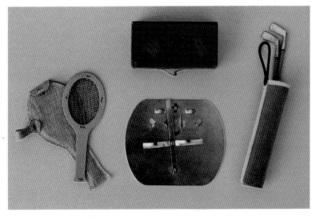

Many Mary Hoyer accessories were offered to complete the doll's costumes. Shown are a tennis racket, suitcase, golf clubs, and a special doll stand (tennis racket $25, suitcase $25-35, golf clubs $25-35, stand $50+). *From the collection of Jan Hershey.*

Monica Doll Studios

The Monica Dolls were marketed by the Monica Doll Studios, located in Hollywood, California, and were created by Mrs. Hansi Share. The first dolls were made in 1941. According to Ursula R. Mertz in her book *Collector's Encyclopedia of American Composition Dolls*, the first dolls came with either composition or cloth bodies. All of the dolls had a very unique feature—instead of using doll wigs as other companies did, human hair was implanted in the composition heads. The Monica dolls were the only composition dolls ever to have rooted hair. All of the dolls had painted eyes and most of them were all composition.

The dolls were very expensive for their time and were usually sold in high end stores. But surprisingly the Montgomery Ward Christmas catalogs also featured several of the Monica dolls in 1946 and 1947. The catalog prices were $22.95 for a 20" doll and 19.95 for a 17" model. The firm also produced 15" and 24" sizes.

The clothes for the Monica dolls were also of high quality. This contributed to the high cost of the dolls.

In 1949, an all plastic doll was added to the line. The new model had moving eyes, unlike the painted eyes on the composition dolls. The dolls continued to be made until approximately 1952.

The Montgomery Ward Christmas catalog featured Monica dolls along with dolls from other firms, including Effanbee, in 1946. The copy for the Monica doll wearing a blue coat and hat read, "The Doll She Dreams About" by Monica of Hollywood. The doll was described as having real human hair "rooted" in her scalp. She was wearing a blue wool felt coat and poke bonnet, along with a dotted swiss dress. She had hand painted eyes, was 17" tall, and sold for $19.95. The Monica doll dressed in a white wool felt coat and hat was 20" tall and sold for $22.95. She also wore a dotted swiss dress. The copy said the eyes were painted by hand.

The Montgomery Ward 1946 Christmas catalog also pictured a 20" Monica dressed in a long blue rayon dress with net trim and a taffeta underskirt. She was priced at $22.95.

Monica Doll Studios

Right:
18" all composition Monica by the Monica Doll Studios. She has hand painted eyes, closed mouth, and real hair rooted in her scalp. She is fully jointed. She is wearing her original glamorous long dress made of blue taffeta, lace, net, and ribbon and comes with her original tag ($400-500).

The 1947 Montgomery Ward Christmas catalog advertised a 20" Monica wearing a net dress with ruffles. The dress had a rayon underskirt, and the doll wore rayon panties, long stockings and slippers. All of these composition Monica dolls had real human hair rooted in the scalp and hand painted eyes. The price for this doll was $22.95

Left:
18" fully jointed composition Monica by the Monica Doll Studios. She has hand painted eyes, closed mouth, and real hair rooted in her scalp. She is in excellent condition and is all original, wearing a long flowered dress trimmed in ribbon. She has her original tag, which reads "MONICA/DOLL/Hollywood." The inside of the tag says "My Dear Doll Mother: You will love to include me into your family, not only because I am beautiful and good, but because you can do so much with me. My real human hair grows right out of my head…it is not a wig glued on…so you can comb and curl it as much as you like without spoiling it. My clothes zip and snap on and off, just like yours, so that you can always keep me in fresh dresses. In any case I will keep you very busy. Monica." ($500-600). *Doll and photograph from the collection of Jo Barckley.*

20" composition fully jointed Monica. She wears her original dress, which has the same trim on the bottom of the underskirt as the doll pictured by Montgomery Ward in 1946. The dress has a rayon net (not crisp like most net) skirt with a heavier rayon underskirt and lace trim. The veil matches the skirt of the dress but it has darkened with age and has several worn places. The flowers on both the veil and the bouquet appear to be original. The underpants are also original and are kept secure with a tie instead of rubber. The doll has had some repainting (200+ in this condition).

Nancy Ann Storybook Dolls, Inc.

Nancy Ann Storybook Dolls, Inc. was incorporated in 1945. It had earlier been founded by Nancy Ann Abbott (real name Rowena Haskins) as the Nancy Ann Dressed Dolls Co. in 1936. Her partner, Les Rowland, joined her in the San Francisco business in 1937.

The first dolls marketed by the Nancy Ann firm were the "Hush-a-Bye Baby" dolls in 1936. The bisque parts to make these dolls were purchased in Japan.

In 1937, little girl models were added to the line. Abbott designed the costumes for these dolls using nursery rhymes or stories for her inspiration. Most of the dolls were 5" tall. The bisque parts for these dolls were also made in Japan.

In 1938, the little girl dolls were marked on their backs "Judy/Ann/USA/Story/Book/USA." These dolls are hard to find and are sought after by today's Nancy Ann collectors.

Other early dolls in demand include the ones with molded socks from the early 1940s. The design of the Nancy Ann dolls changed to a stiff leg model circa 1943, when the head, body, and legs were made in one piece. Only the arms were jointed. This design continued to be used until 1947. At that time, the Nancy Ann Co. joined other doll manufacturers in producing dolls made of plastic.

The Nancy Ann Storybook firm operated its own potteries during the World War II years. One factory was located in Berkeley, California and another was opened in Stockton, California in 1944. Besides dolls, the company made dishes for Navy hospitals during the war years.

The Nancy Ann painted over bisque baby dolls came in various sizes during the years, ranging from 3-1/2" to 4-1/2". The molds used to make the baby's hands also changed from star-shaped to closed fists. The painted over bisque little girl models (also used as boys with male clothing added) ranged in size from 4-1/2" to 7". The earlier dolls had jointed arms and legs while the later dolls had stiff legs.

Painted over bisque dolls were issued in a variety of costumes under different series names. The same series titles were used for several years and costume designs were changed frequently, so several different designs could carry the same number and name. Dolls were produced that were named after the days of the week, the months of the year, the seasons, nursery rhymes, fairy tales, book characters, various countries, sports, operettas, and famous songs. The dolls in the "Family Series" were especially popular because these dolls included a wedding party. Dolls were dressed as a bride, groom, bridesmaid, flower girl, and ring bearer.

Besides dolls, the Nancy Ann Company also produced furniture for its dolls. These items were mostly made of cardboard and printed taffeta. Furniture pieces included a bed, chaise, dressing table, and easy chair. The furniture was only made for a short time. McCall also marketed a pattern so the consumer could make this same type of furniture at home. It was #811 and was issued circa 1940. Patterns were included for a baby bassinet, dressing table and stool, easy chair and ottoman, couch, and bed. The furniture was designed to fit dolls 4" to 7" tall.

Most of the current Nancy Ann Storybook collectors prefer mint, tagged dolls, in their original boxes. The value of the dolls is in their costumes, tags, and boxes. Sometimes, having the tag or the marked box is the only way a doll can be identified. The older dolls with jointed legs are the most desirable and, therefore, the most expensive.

The plastic Nancy Ann dolls continued to be sold through the later 1940s and into the 1950s. With the changing doll market and the popularity of the Vogue Ginger dolls, the Nancy Ann firm developed a new similar doll, called Muffie, which was also a success. Other new dolls followed, including Nancy Ann Style Show, Debbie, Lori-Ann, Miss Nancy Ann, Little Miss Nancy Ann, and Baby Sue Sue (see *Dolls and Accessories of the 1950s* for more information about these dolls).

As the decade of the 1950s drew to a close, so did the success of the Nancy Ann Storybook Company. According to Marjorie A. Miller, writing in her book *Nancy Ann Storybook Dolls*, the business moved to smaller quarters in San Francisco in 1961 as its owners experienced health problems. Nancy Ann died in 1965 and Lee moved the company into bankruptcy. Eventually the assets were purchased by Albert M. Bourla and he brought out a new "Muffie Around the World" series in 1967.

Despite the hardships endured by Nancy Ann Abbott during the last years of her life, collectors will continue to remember her because of the beautiful doll clothing she designed. The products of Nancy Ann Storybook Dolls, Inc. remain near the top in their popularity with today's collectors.

3-1/2" Nancy Ann Rock-a-Bye Baby Series #211, circa 1938. The painted over bisque doll has painted eyes, closed mouth, molded hair, and curved arms. She is marked on the back "87 Made in Japan." She is wearing a long dress and bonnet and came in a Sunburst Box. The sticker reads "STORYBOOK DOLLS." (MIB $500+). *Doll and photograph from the collection of Nancy Roeder.*

Nancy Ann Storybook Dolls, Inc.

Nancy Ann painted over bisque babies, circa 1930s and early 1940. The dolls on the far left and right in the front row are 4-1/2" tall and have "star hands." The babies in the center front are "Japan" mold dolls and are 3-1/2" tall. The back row of dolls are, left to right: #210 Hush-a-Bye Baby Series (long dress), circa 1938; #201 Hush-a-Bye Baby Series (short dress and bonnet), circa 1940; #201 Hush-a-bye Baby Series (short dress and bonnet), circa 1941. All of these dolls have the "star" hands ($250 each, more for earliest and boxed). *Dolls and photograph from the collection of Nancy Roeder.*

Nancy Ann early Bo Peep dolls, circa 1937-38. All three dolls are 5" tall with jointed arms and legs. Bo Peeps on the left and center are marked "Made in Japan 1146." The center doll also has her original sticker, which reads "Nancy Ann Dressed Dolls." The doll on the right is marked "Japan." ($500+ without damage). *Dolls and photograph from the collection of Nancy Roeder.*

5" Scotch dolls from Around the World Series, circa 1937-39. All the dolls have jointed arms and legs. The doll on the far left is marked "Made in Japan 1146" ($500+). The doll second from left is marked "Judy Ann" (boxed $1,600+) and the doll third from left is marked "America" (boxed $1,000+). The doll on the far right has molded socks and also molded bangs under her wig ($400+). *Dolls and photograph from the collection of Nancy Roeder.*

Nancy Ann Storybook Dolls, Inc. 101

Four Nancy Ann painted over bisque dolls showing different boxes. On the left is Bo Peep #21, Made in Japan 1146, circa 1937-38 with a Sunburst Box (MIB $650). Second from left is #89 Mammy, circa 1939, with molded socks and molded bangs, in a red box with silver dots and silver label (MIB $600). Third from left is He Loves Me #121, circa 1940. She came in a pink box with large white dots (MIB $300). On the right is I Have a Little Pet, circa 1942, which came with a cat. She has a white box with blue dots and a silver label (MIB with pet $400+). *Dolls and photograph from the collection of Nancy Roeder.*

5" Judy Ann in Storybook #300 (packaged to look like a book), circa 1939. Also called Judy Ann in Fairyland. The doll has molded socks and molded bangs under her wig as well as jointed legs. The set included three additional sets of clothing (MIB complete $5,000). *Doll and photograph from the collection of Nancy Roeder.*

5" To Market #12, circa 1938. She is marked "Judy Ann" and is very unusual because she has painted molded hair. She is missing her small black purse. She came with the early Sunburst box with a gold label (boxed $1,200+). *Doll and photograph from the collection of Nancy Roeder.*

5" Hansel & Grethel (sic) #177, circa 1939. Both dolls have molded socks, molded bangs under their wigs, and jointed legs. They were packaged together in a blue box with small silver dots. Both dolls have their original tags, which read "STORYBOOK DOLLS." (MIB $1,200+). *Dolls and photograph from the collection of Nancy Roeder.*

5" Nancy Ann dolls from the Masquerade Series. From left: Cowboy #62 or 64 with molded socks and molded bangs, circa 1939 ($1,500); Pirate #61 with molded socks, circa 1940 (MIB $1,000); Ballet Dancer #63 with molded socks and molded bangs, circa 1939 (MIB $1,800); Gypsy #60 with molded socks, circa 1940 ($1,000). *Dolls and photograph from the collection of Nancy Roeder.*

5" Around the World Series: Chinese #33, circa 1940. She has molded socks, is jointed bisque, has painted hair, and different painted features to give her an oriental look. Her original tag reads "STORYBOOK DOLLS" ($900+). *Doll and photograph from the collection of Nancy Roeder.*

Left:
5" Sports Series: Skiing #73, circa 1940. She has molded socks and white boots ($1,200). *Doll and photograph from the collection of Nancy Roeder.*

5" Sports Series: Riding #72, circa 1939. She has molded socks, molded bangs, and black boots. She comes with her original red box with silver dots and silver label (boxed $1,000). *Doll and photograph from the collection of Nancy Roeder.*

Right:
5" Flower Series: Black-eyed Susan #4, circa 1940. She has unusual brown eyes and molded socks. She comes with her original tag and pink box with white polka dots (MIB $600). *Doll and photograph from the collection of Nancy Roeder.*

5" doll in Boudoir Box #900, circa 1940. Original box is 18" x 18". The set included the doll, bed, chaise and pillow, dressing table, stool, and metal mirror. The doll has molded socks and jointed legs (MIB $1,800). *Doll and photograph from the collection of Nancy Roeder.*

5-1/2" Margie Ann in Hat and Coat #82, circa 1941. She is pudgy with white boots, jointed legs, and wrist tag. Her box is white with blue dots and silver label. She comes from the Family Series (MIB $450). *Doll and photograph from the collection of Nancy Roeder.*

5-1/2" Topsy & Eva dolls #176, circa 1941. They were sold as a pair in one box. Pudgy, jointed leg Topsy wears black boots with white buttons (MIB $1,000). *Dolls and photograph from the collection of Nancy Roeder.*

5-1/2" Pussy Cat, Pussy Cat #126, circa 1941. She is the pudgy model with jointed legs and white boots. She includes her original pipe cleaner cat ($450 original with cat). *Doll and photograph from the collection of Nancy Roeder.*

Nancy Ann Storybook Dolls, Inc.

Right:
6-1/2" December #198, circa 1941. She is bisque with socket head and jointed legs. A drawing of this doll was used by the Nancy Ann firm as a "logo" on their brochures. She is all original with her wrist tag, brochure, and box (MIB $200-225). *Doll and photograph from the collection of Nancy Roeder.*

Below:
7" In Powder and Crinoline Series: Lady in Waiting, Charmaine #253, circa 1942. She is bisque with jointed legs. The whole series of dolls included the Prince, Princess, and ten Ladies in Waiting (MIB $200). *Doll and photograph from the collection of Nancy Roeder.*

Right:
5-1/2" Mistress Mary #119, circa 1944-45. The bisque doll has stiff legs and is all original with her wrist tag and box (MIB $75+). *Doll and photograph from the collection of Nancy Roeder.*

Left:
G. Fox & Co. advertised Nancy Ann dolls in their catalog in December 1942. Included were Dolls of the Week, Dolls of the Month, Storybook Dolls, and Romantic Dolls. The dolls sold from 75 cents to $2.29 each. *From the collection of Marge Meisinger.*

Nancy Ann Storybook Dolls, Inc. 105

Left:
5-1/2" Family Series Bride #86, circa mid 1940s. The bisque doll has stiff legs and is all original with her wrist tag and box. These dolls were marked on the back, "Storybook Doll USA." Her wrist tag reads "Storybook Dolls by Nancy Ann/Family Series" (MIB $75).

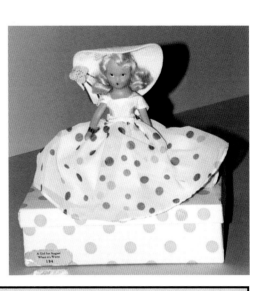

Right:
6-1/2" August #194, circa 1945-46. This is another later bisque doll with the stiff legs. She is all original with her wrist tag and box (MIB $100). *Doll and photograph from the collection of Nancy Roeder.*

5-1/2" Queen of Hearts #157, circa mid 1940s. She has stiff legs and is all original with her wrist tag and box. She has an added heart insert which makes her a little more special (MIB $75+).

NANCY ANN STORYBOOK DOLL IDENTIFICATION

DATE	GIRL MOLDS	BABY MOLDS	BOX	BOX LABEL	FASTENER	IDENTIFICATION	BROCHURE
c.1936	none	bisque "87 Made in Japan" "88 Made in Japan" "93 Made in Japan" other Japanese molds	pink or blue mottled or marbleized box; colored box with sunburst pattern	gold	silver pin	gold foil sticker on clothes "Nancy Ann Dressed Dolls"	none
c.1937	bisque "Made in Japan 1146" "Made in Japan 1148" "Japan" other Japanese molds	bisque marks as above	colored box with sunburst pattern	gold	silver pin	gold foil sticker on clothes "Nancy Ann Dressed Dolls"	none
c.1938	bisque "America"	bisque marks as above	as above	gold	silver pin	gold foil sticker on clothes "Judy Ann"	none
c.1938	bisque "JUDY ANN USA" "STORY BOOK USA"	bisque marks as above plus "STORY BOOK USA"	as above	gold transition to silver	brass pin	gold foil sticker on clothes "Storybook Dolls"	none
c.1939	bisque molded socks & molded bangs	bisque star shaped hands	colored box with small silver dots	silver	brass pin	gold foil sticker on clothes "Storybook Dolls"	none
c.1940	bisque molded socks (only)	bisque star shaped hands	colored box with white polka dots	silver	brass pin	gold foil sticker on clothes "Storybook Dolls"	yes
c.1941-42	bisque pudgy tummy or slim tummy	bisque star hands transition to fist	white box with colored dots	silver	brass pin	gold foil bracelet with name of doll	yes
c.1943-47	bisque stiff legs (head, body & legs all in one piece)	bisque fist hands	as above	silver	ribbon tie, dull silver pin, or brass pin	gold foil bracelet with name of doll	yes
c.1947-49	plastic painted eyes	bisque body plastic arms & legs	as above with "Nancy Ann Storybook Dolls" between dots	silver	brass snap	gold foil bracket with name of doll	yes
c.1949(?)	plastic black sleep eyes	plastic black sleep eyes	as above	silver	brass or painted snap	gold foil bracelet with name of doll	yes
c.1953(?)	plastic blue sleep eyes except 4 1/2" girls	plastic black sleep eyes	as above often with clear lids	silver	large gripper "doughnut" snap	gold foil bracelet with name of doll	yes

Prepared by: Nancy Roeder and Mary Lu Trowbridge

Nancy Ann Storybook Doll Identification Chart, compiled by Nancy Roeder and Mary Lu Trowbridge. It shows dates, boxes, molds, fasteners etc. for these popular dolls. *Courtesy of Nancy Roeder and Mary Lu Trowbridge.*

Left:
6-1/2" Operetta Series: Naughty Marietta #307, circa 1946-47. This series was marketed in bisque for a very short time. The doll is made in the style of the last of the bisque dolls with a stiff bisque body and plastic arms. She is all original with her wrist tag and box (MIB $135-145). *Doll and photograph from the collection of Nancy Roeder.*

Vogue Dolls, Inc.

Jennie Graves, of Somerville, Massachusetts, began her doll business in the 1920s when she dressed dolls to sell to a Boston department store. Graves used the name Vogue Doll Shoppe. When the business grew, Graves hired home sewers to help meet the demand. One of the most popular of the early dolls dressed by Vogue was the "Just Me" German bisque doll. These dolls ranged in size from approximately 7" to 10". They featured bisque heads and composition bodies and were made by the German Armand Marseille Doll Co. The Vogue firm also dressed a variety of other celluloid, bisque, and rubber dolls.

Jennie's husband died in 1939 and she relocated to Medford, Massachusetts, where she continued to dress dolls made by other manufacturers.

During the 1930s, composition dolls were used by the Vogue firm as the basis for their business. The basic dolls were purchased from other firms, including Ideal and Arranbee. Baby dolls with composition heads, arms and legs, and cloth bodies, as well as all composition little girl dolls, were dressed by Vogue.

One of the most collectible of these little girl dolls is the model called Dora Lee. Like nearly all of the early Vogue dolls, these dolls were unmarked except for a Vogue tag. The dolls were 11" tall and featured sleep eyes and a closed mouth. They had molded hair under their wigs. Some of the dolls had "Dora Lee" stamped on the bottom of their shoes. The dolls were sold by Vogue from the late 1930s through the early 1940s.

The famous Vogue Toddles all composition dolls were first marketed by the company in 1937. At first, these 8" dolls were purchased from the Arranbee Doll Co., then dressed and marketed by Vogue. Later, Mrs. Graves had Bernard Lipfert design a similar doll just for Vogue. The markings on these dolls changed through the years. From 1937 to the early 1940s, the dolls were marked "R&B/Doll Co." on their backs. From 1942 to 1943 some of the dolls were simply marked "Doll Co." Other markings used from 1940-43 were "Vogue" on the head and "Doll Co." on the back. The later composition dolls were marked "Vogue" on both the head and the back.

The dolls have mohair wigs, painted eyes, and closed mouths. They were sold until 1948 when the new hard plastic doll, eventually known as Ginny, took over the small doll Vogue market.

The Toddles dolls were issued in a variety of costumes, including those representing characters in nursery rhymes and fairy tales. Other dolls were dressed in outfits depicting the native dress of foreign countries. Some of the most popular Toddles dolls are those relating to World War II (see Military Dolls chapter). Included were dolls representing the various branches of military services plus "home front" dolls such as an Air Raid Warden. A different line featured patriotic dolls like Uncle Sam and Miss America.

Another small, all composition doll that sold for many years was the Sunshine Baby. These 8" baby dolls were marketed from the late 1930s until the mid 1940s. The dolls featured the same heads and bodies as those used on the Toddles dolls, but by adding curved arms and legs the dolls were made into infant models. The babies were marked "Vogue" on the back of the heads and "Doll Co." on their backs. They featured painted eyes and molded hair. Many of the outfits were fastened with string ties.

During the war years, Vogue joined many other companies in producing dolls dressed in uniform. Besides the special Toddles dolls, Vogue also issued WAVE-ette and WAAC-ette dolls. These all composition little girl dolls were dressed in uniforms representing the W.A.V.E and W.A.A.C women's branches of the Army and Navy. They were 13" tall and the only marking was a Vogue sticker tag. They had sleep eyes and wigs.

Another series of Vogue dolls, particularly attractive for collectors, are the "Make-Up-Dolls." These composition dolls date from the 1940s and came in three sizes: 13" (Mary Jane), 16" (Betty Jane), and 19-1/2" (Jennie). All the dolls were from the regular line but along with attractive clothing, each doll was supplied with a "make-up kit." These kits consisted of make-up accessories, including a mirror, powder can, and powder puff. The dolls' dresses were tagged "Vogue Dolls, Inc./Medford, Mass."

According to information in the book *Collector's Encyclopedia of Vogue Dolls*, by Judith Izen and Carol Stover, Vogue marketed a very creative women's sports series of dolls in the early 1940s. These very rare 16" composition dolls were costumed in various sports attire. Besides the outfits, the dolls were supplied with accessories appropriate to their costumes. Included were ice skates, skis, tennis rackets, and golf clubs. The dolls had sleep eyes, real lashes, closed mouths, and wigs.

As the composition era was winding down, Vogue advertised a series of dolls wearing pantalettes. This line dates from 1947-48. The dolls were 13" tall with sleep eyes and mohair wigs. All were very nicely dressed in old-fashioned costumes. The Montgomery Ward Christmas catalog for 1947 pictured one of these dolls, called a Pantalette Doll, for $6.95.

As the end of the decade of the 1940s approached, Vogue, like the other major doll firms, began to produce dolls from hard plastic. Dolls included 14" little girl models as well as the very collectible Crib Crowd baby series, which began in 1948. The 8" painted eye, all plastic forerunner to the 8" Ginny doll also made its appearance in 1948. With these new models, Vogue Dolls, Inc. was soon to experience success as never before.

Collectors still treasure the beautifully dressed composition dolls that were marketed by the firm in the 1930s and 1940s. Since Vogue purchased many of their basic dolls from other companies, most of these dolls were marked only with a Vogue sticker or a dress tag. For that reason, it is hard to identify many of the dolls. The *Collector's Encyclopedia of Vogue Dolls* offers hundreds of pictures to help collectors identify their Vogue dolls. It is highly recommended.

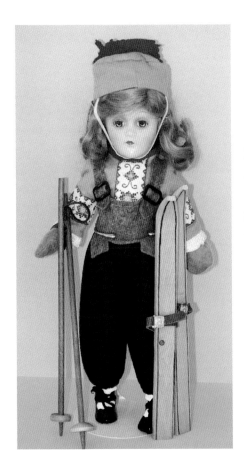

Left:
11" composition doll thought to be Dora Lee by Vogue. She has sleep eyes, closed mouth, mohair wig over molded hair, and is fully jointed. Some of these dolls were stamped with the Dora Lee name on the bottom of their shoes but both outer soles are missing on her shoes. She is all original, complete with skis. The skis appear to have been made by the same maker who produced the skis sold with the Vogue Sports Women series of 16" composition dolls ($300+ unmarked and no tag).

Right:
8" composition Vogue Toddles, circa 1943. She has painted eyes, closed mouth, mohair wig, and is fully jointed. She is marked "DOLL CO." on her back. She is all original, wearing an untagged dress with underpants attached to the dress. Ties in the back keep the dress closed. She also wears a straw hat, leatherette shoes, and socks. Her hair may have been trimmed and she has paint flaking ($100).

8" composition Vogue Toddles, circa 1943. She has painted eyes, closed mouth, mohair wig, and is fully jointed. The doll has a bent right arm and is marked "DOLL CO." on the back. She is all original wearing an untagged yellow organdy dress, attached slip and panties, yellow straw hat, and leatherette shoes ($250+). *Doll and photograph from the collection of Carol Stover.*

8" composition Vogue Toddles, circa 1946. She has painted eyes, closed mouth, mohair wig, and is fully jointed. She has a bent right elbow and is marked "DOLL CO." on her back. She is wearing an untagged long pink organdy print dress with attached long pantalettes, tie-on white apron, and matching pink cap ($250). *Doll and photograph from the collection of Carol Stover.*

8" composition Vogue Toddles, circa mid to late 1940s. She has painted eyes, closed mouth, mohair wig, and is fully jointed. She has two straight arms and is marked "VOGUE" on her head and body. She wears a blue organdy dress with a matching bonnet, and slip-on white leatherette shoes. Her dress is tagged with a Vogue "inkspot" tag. She is pictured with her original box ($350 with box). *Doll and photograph from the collection of Carol Stover.*

8" composition Vogue Toddles Bunky and Binky twins, circa 1947-48. They have painted eyes, closed mouths, mohair wigs, and are fully jointed. The dolls are marked "VOGUE" on their heads and backs. They are from the Playmates Series and are dressed in matching blue jersey shirts and bottoms. The clothing is marked with the "inkspot" tag. They are missing their hats ($100-125 each with some paint flecking and hair wear).

Below:
8" composition Vogue Sunshine Babies, circa late 1930s to mid 1940s. They have painted eyes, molded hair and are fully jointed. They are marked "Vogue" on the back of their necks. Both dolls are wearing their original dresses and matching bonnets ($300-350+ each). *Dolls and photograph from the collection of Carol Stover.*

Left:
8" composition Vogue Toddles, circa 1948. She has painted side-glancing eyes, closed mouth, mohair wig, and is fully jointed. She wears a checked sun suit, straw sun hat, and replaced shoes ($200). *Doll and photograph from the collection of Carol Stover.*

Right:
14" composition unmarked Vogue Betty Jane, circa 1943. She has sleep eyes, closed mouth, mohair wig in braids, and is fully jointed. She is all original wearing a pinafore dress displayed at the Toy Fair circa 1942-43 ($325-350+). *Doll and photograph from the collection of Carol Stover.*

13" composition Vogue unmarked doll, circa 1940s. She has painted eyes, mohair wig, closed mouth, and a bent elbow. The doll is wearing a brown dress trimmed with colored felt flowers, attached panties, and original tie shoes and socks. She is probably missing a hat or ribbon ($200-250). *Doll and photograph from the collection of Carol Stover.*

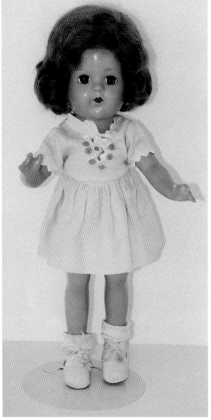

13" composition Mary Jane, probably made by Ideal but dressed by Vogue. She has sleep eyes, closed mouth, mohair wig, and is fully jointed. She is marked 13". She is wearing a yellow dress trimmed with felt flowers and a tie at the neck. Her tie shoes and socks are also original ($200-250).

Right:
13" composition Make-Up Doll, circa 1940s. She has sleep eyes, closed mouth, mohair wig, and is fully jointed. She is missing her drawstring makeup bag. She is wearing her original white organdy dress, matching bonnet, underwear, and shoes and socks. The original dress tag read "Vogue Dolls, Inc./Medford, Mass." in blue ink on a white tag ($350+). *Doll and photograph from the collection of Carol Stover.*

13" and 18" unmarked Vogue dolls, circa 1947-48. The smaller dolls have sleep eyes, closed mouths, mohair wigs, and are fully jointed. The larger doll has sleep eyes, an open mouth with teeth, a mohair wig, and is also fully jointed. The dolls are all original wearing tagged Vogue dresses. The doll in the long dress has a one strip tag while the other two dolls have tags that are printed on both sides. All of the tags have blue backgrounds with "Vogue Dolls" printed in white in large letters and "Medford, Mass." printed in smaller letters. The 18" doll wears a skirt and blouse with ruffled eyelet trim and ribbon bows. She is Polly from the Junior Miss series from 1948. The doll on the right wears a plaid jumper and beret and is the doll called Ginger in the Young Folks series from the Vogue price list for 1948. The doll on the left is dressed in a long gown with rickrack trim, a matching hat, and long pantalettes. This doll is Joan, one of the Young Folks dolls from 1947 (13" with tagged dresses $375-425, 18" with tagged dress $500+). *Dolls from the collection of Marge Meisinger. Photograph by Carol Stover.*

13" Vogue "Pantalette" doll as pictured in the 1947 Montgomery Ward catalog. She sold for $6.75.

13" Vogue composition "Pantalette" doll circa 1947. She has sleep eyes, closed mouth, a mohair wig, and is fully jointed. She is all original wearing an organdy skirt over lace-trimmed pantalettes and petticoat. She also wears a felt jacket, and a poke bonnet. The tag for this dress featured a blue background with "Vogue Dolls/ Medford, Mass." in white. A price list pictured in the *Collector's Encyclopedia of Vogue Dolls* by Izen and Stover identifies this doll as Jean from the Young Folks series of 1947 ($375-425). *Doll and photograph from the collection of Carol Stover.*

Miscellaneous Dolls

Boudoir Dolls

After 1900 and until the early 1930s, women's magazines were filled with patterns to make embroidered gift items. As the years passed, these patterns changed from linens and children's clothing to all kinds of knick knacks for the "modern woman." These included: draw string bags, bedroom slippers, cases for handkerchiefs, holders for dirty laundry, and fancy pillows. Finally, these useful gifts gave way to items strictly for display, such as lamps, pincushions, and bed dolls.

These dolls were not meant as toys for little girls but were to be displayed in the bedroom of an adult woman. The early dolls were made mainly of cloth and resembled the 1920s female flapper with extra long arms and legs. They usually had costumes that were separate from the doll.

For the women who weren't able to sew, many companies manufactured similar dolls. The Unique Novelty Doll Co. and the Sterling Doll Co., both from New York, advertised this type of doll as late as 1932. The faces of the commercial dolls were made of cloth stiffened with starch. The more expensive dolls had shaped cloth faces of buckram, with fancy wigs of mohair. The Gerling Toy Company of New York offered similar dolls with a composition head during the late 1920s and early 1930s.

The doll that is known as the boudoir doll is a later version of the flapper doll and it is not as thin. These dolls were usually from 24" to 36" tall.

Some of the boudoir dolls were completely made of cloth, with elaborate dresses made of satin or organdy trimmed with lace and ruffles to make the doll look very romantic. These boudoir dolls were also not made for children, but rather for women to use in decorating—to give a touch of glamour to their bedrooms.

During the 1930s, similar dolls were made with a composition head, lower arms and legs, and cloth bodies. These dolls were marketed to tie in with the glamour promoted by motion pictures and many of the dolls were highly painted with lots of eye shadow and other make-up. Some had added eyelashes and a number of them were quite buxom. Many of these dolls wore molded high heel shoes and nearly all were dressed in long dresses. Wigs were usually of mohair.

Their cloth bodies were sometimes filled with excelsior or cheap dark colored batting. The boudoir doll was popular for many years and was still being sold during World War II by well-known candy stores. Many servicemen gave a gift of one of these dolls to a current girlfriend.

For collectors, the most sought after boudoir doll is the doll smoking a cigarette. These were made in the late 1920s and the early 1930s. Some of them had the look of Marlene Dietrich, as they were dressed in pants and wore a top hat as Marlene so often did.

During later years, a boudoir doll was produced which had a vinyl head but it did not prove very popular and was not made for any great length of time.

32" all cloth Boudoir doll, circa early 1930s. Her face is made of molded twill material with painted features. She has a mohair wig, long arms and legs, and is stuffed with cotton. She is wearing her original removable clothing and shoes ($75+).

N. Shure Co. of Chicago advertised Boudoir dolls in their catalog for 1938. Six different models were pictured, ranging in size from 24" to 31". The dolls had composition heads, arms, and feet, and mohair wigs. Each was dressed in a long dress with a full skirt. *From the collection of Marge Meisinger.*

Miscellaneous Dolls

26" Boudoir doll, circa late 1930s to early 1940s, with a composition head, shoulder plate, arms, and lower legs. She has painted eyes, a closed mouth, and a mohair wig. Her bust is cotton stuffed. The original white satin costume appears to be a bridal dress ($75-100).

27" Boudoir doll, circa late 1930s to early 1940s. She has painted features, a mohair wig, composition head, lower arms and legs. She is all original, wearing a satin dress trimmed with lace ($75+). *Courtesy of JJs Doll Company.*

25" doll with composition head and shoulder plate ending at the waist. She has a molded bust and composition arms that extend to the shoulders. Her features are painted with added eyelashes. She has a cloth body and composition feet with painted high heel shoes. Her wig is mohair and her Halloween costume is all original ($125+).

22" "Smoking" doll, circa early 1930s. She has painted features, a mohair wig, and composition lower arms and legs. The rest of the doll is cloth except for her face, which is molded from a heavy material that has a felt-like finish. She is all original and her costume includes a top hat. The doll has the look of Marlene Dietrich in her early films. The cigarette in her mouth makes this doll most desirable ($300+).

Miscellaneous Dolls 113

Two very similar 29" cloth Boudoir dolls with faces molded of a heavy material, painted features, and added eyelashes. Both have mohair wigs and removable shoes. The doll on the left has lower arms of composition. Her original costume is made of organdy. The original clothing of the doll on the right is made of net, lace, and taffeta. Both dolls date from the 1930s. They are displayed on a bedspread of the period against hand made pillows also of the era ($100-150 each).

Bye-Lo Baby

The Bye-Lo Baby is one of the truly unique dolls of the twentieth century. It was designed by Grace S. Putnam and was first issued in 1923. The doll had been modeled from a newborn infant. The Bye-Lo dolls from the early 1920s featured bisque heads and cloth bodies but by 1927 the dolls were also being marketed with heads made of composition. Jan Foulke, writing in the *14th Blue Book*, listed the Cameo Doll Co. as the maker of the composition doll heads. The dolls with composition heads were still being advertised in 1934, 1937, and 1943 in various catalogs and magazines. It may be that the dolls were made continuously during that time. The dolls in these later ads came in sizes of 13", 15-1/2", and 18".

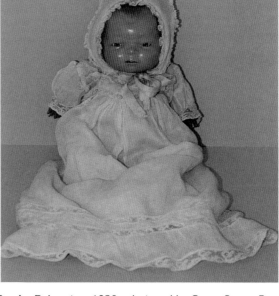

12" Bye-Lo Baby, circa 1930s, designed by Grace Storey Putnam. She has a composition flange head, celluloid hands, and a cloth body and arms with curved legs. She has sleep eyes, closed mouth, and molded hair. She has a crier that still works. The head and body are marked "Grace Storey/Putnam." She has been redressed ($200).

The Bye-Lo baby was still being advertised in the John Plain catalog in 1937. The doll came in three sizes: 13", 15-1/2", and 18". She was modeled from a three-day old baby by designer Grace Storey Putnam. *From the collection of Marge Meisinger.*

Freundlich, Ralph A.

Ralph A. Freundlich started his business in 1929. In 1934, the factory operations were moved to Clinton, Massachusetts.

The firm manufactured a number of different types of dolls during the 1930s and 1940s. Although most of these dolls were made to sell very cheaply, the firm made special dolls that are of real interest to today's collector. These included dolls representing Little Orphan Annie (see Character Dolls chapter), General MacArthur and Baby Sandy (see Personality Dolls chapter), and a series of World War II military dolls (see Military Dolls chapter).

Besides these unique dolls, the firm also produced a Red Riding Hood, Grandma, and Wolf set of dolls in 1934. These 10" dolls were advertised in the Sears catalog for that year and were priced at $1.00 for the set. The firm also marketed a similar set of 10" dolls in the images of the "Three Little Pigs and the Wolf."

Another interesting set of dolls made by Freundlich was the unauthorized set of 7" composition Quintuplets (because of the popularity of the Dionne Quints) and a 9" nurse. The dolls were packaged in a suitcase and came complete with a layette. They had the look of the Alexander small baby Quints but they were made more cheaply.

Cloth dolls with molded hair, composition heads, and "Goo Goo" eyes were also a part of the Freundlich line for several years. These dolls came in a variety of models and sizes.

Since most of the Freundlich dolls were not marked, it is hard to identify them unless they have an original tag or are dolls known to have been produced by the company. The dolls were made to sell cheaply and many were carried by the "dime" stores of the period.

According to Marian H. Schmuhl, in a February 1994 *Doll Reader* article, the company ceased doing business in 1945.

17" cloth doll with composition head, attributed to Ralph A. Freundlich, circa early to mid-1940s. She has "goo goo" eyes, molded hair, and a closed mouth. The rest of the doll is stuffed cloth. A smaller version of this head was used on a doll with a printed cloth body and wooden "Dutch" shoes, circa 1943.

Gem Toy Company

The Gem Toy Company, based in New York, was in business from the early teens until the 1930s. The firm made many different dolls over several decades but the dolls were usually very similar to dolls made by other firms. Some of them were marked "Gem" on the backs of the necks but more often only a tag on the clothing can be used to identify the dolls from this company.

Dolls that are known to have been made by Gem include "Mama" dolls, a black composition Topsy doll, and all composition little girl dolls including a Patsy look-a-like advertised in 1931.

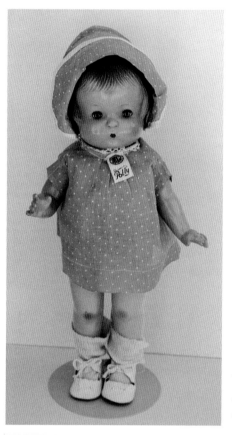

14" composition Polly by Gem Toy Co., circa early 1930s. She has tin sleep eyes, closed mouth, molded hair, and is fully jointed. Her original dress is tagged "Gem/Trademark Reg./Made in U.S.A./Polly." ($150-200). *From the collection of Jan Hershey.*

Goldberger

Eugene Goldberger's company was established in Brooklyn, New York in 1917. The firm used the trademark "EEGEE" beginning in 1923. The trade name came from the founder's name, E. Goldberger. Most of the company's dolls were of lower quality and cost but two dolls from the 1930s are very popular with Shirley Temple collectors. These include Little Miss Movie and Little Miss Charming, both look-a-like Shirley Temple dolls. Miss Charming came in sizes from 12" to 27". In 1936, Sears advertised the doll as a walking doll priced at $2.98 for the 16" model and $4.95 for the larger 20" doll. The all composition dolls had sleep eyes and mohair wigs.

Besides these special dolls, the company produced composition baby dolls (including a Baby Charming), little girl dolls (including a Patsy look-a-like), and other dolls similar to those from other lesser known companies. The Baby Charming dolls resembled the Alexander Dionne Quint toddler dolls, having black molded hair and sleep eyes. Since most of the Goldberger dolls are not marked, they are hard to identify without the tags, boxes, or advertising.

The firm was still producing dolls in the 1950s of both hard plastic and vinyl.

Miscellaneous Dolls 115

Left:
19" Little Miss Movie composition doll by EEGEE (Goldberger), circa 1936. She has tin sleep eyes, open mouth with teeth, and a blonde mohair wig. She wears her original Goldberger dress copied from the movie *Bright Eyes*. She still has her original pin, which reads "EVERYBODY LOVES ME LITTLE MISS MOVIE." This doll was made to compete with Ideal's Shirley Temple dolls ($450-550). *From the collection of Marge Meisinger. Photograph by Carol Stover.*

J. Halpern Co. (Halco)

The J. Halpern Co., of Pittsburgh, Pennsylvania marketed nice composition dolls that sold for reasonable prices during the 1940s. Their trade name was Halco and their logo was "Superb Halco Brand Quality Dolls." Polly and Pam Judd mention the firm in their *Hard Plastic Dolls* book and show ads of their hard plastic dolls in the early 1950s. One of these dolls was a "Miss Fluffee" doll, 24" tall with a hard plastic head, cloth body, and latex limbs. A "Fluffee" doll with a composition head, limbs, and a cloth body was also made by the firm.

The Sears catalog for 1936 advertised the Miss Charming Walking doll made by Goldberger. The composition doll had sleep eyes, open mouth with teeth, mohair wig, and was fully jointed. She came in a 16" size for $2.98 and a 20" size for $4.98.

18" composition Baby Fluffee doll made by the J. Halpern Co., circa 1946. She has flirty sleep eyes, closed mouth, a mohair wig, composition lower arms and legs, a cloth body (containing a crier) and cloth upper arms and legs. The doll is all original except for her shoes and socks. She wears a dotted swiss dress and coat and hat. She still has her original tag, which reads "Superb Halco Brand/ Baby/ Fluffee." ($125-150). *Childhood doll of Bonnie McCullough. Photograph by Marilyn Pittman.*

Hedwig Dolls

The Hedwig Dolls were all composition dolls dressed in costumes that came from characters in the books of Marquerite de Angeli. The 14" dolls date from the 1940s and were dressed by Hedwig of Philadelphia. The original wrist tags read "Hedwig Dolls/ Registered Authorized/from the books of Marquerite de Angeli." The dolls had sleep eyes or painted eyes, closed mouths, and mohair wigs. They included the following characters: Hannah from *Thee Hannah* (Quaker costume), Lydia from *Henner's Lydia* (Pennsylvania Dutch), Suzanne from *Petite Suzanne* (French Canadian), and Elin from *Elin's Amerika* (Swedish costume).

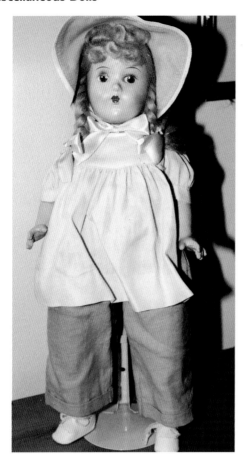

20" mint Halco doll made by the J. Halpern Co., circa 1940s. She has painted eyes, closed mouth, mohair wig, composition arms and legs, and a cloth body. The doll is all original and comes with her box ($200-250). *Doll and photograph from the collection of Veronica Jochens.*

13" composition Suzanne by Hedwig of Philadelphia. She has sleep eyes, closed mouth, blonde mohair wig, and is fully jointed. She is all original with her wrist tag. She is wearing a cotton print dress, slip with attached panty, red coat and cap, and black socks and leatherette shoes. Her tag reads "HEDWIG DOLLS/ Registered Authorized/From the Books of/Marguerite de Angeli." ($550+). *Courtesy of Frashers' Doll Auctions, Inc.*

The box for the 20" Halco doll says "SUPERB/ HALCO/ BRAND/ QUALITY DOLLS." *Box and photograph from the collection of Veronica Jochens.*

Hollywood Doll Manufacturing Co.

The Hollywood Doll Manufacturing Co., of Glendale, California, produced small composition dolls during the 1940s. Most of the identified dolls are 5" or 9" tall. The dolls competed with the Nancy Ann Storybook dolls but were of lesser quality. A basic doll was used and the identity was changed by dressing the doll in different costumes. There were several lines of dolls made in each size. The 5" dolls came in the Princess Series (six dolls), Lucky

Star Series (twelve dolls, one for each Zodiac sign), Hollywood Book Series (six dolls), Nursery Rhymes (six dolls), Playmates (six dolls), and Little Friends (six dolls). The 9" dolls came dressed as a bride, as a cowboy and cowgirl in the Western series, and as Punchinello, a clown. There was also an Idyll Series of six dolls, a Garden Series of six dolls, and a Toyland Series of twelve dolls.

The dolls were made of composition with painted eyes, mohair wigs, and jointed arms and legs. The firm also issued the same type of dolls in hard plastic when composition dolls lost favor.

Left:
5" composition Lucky Star doll made by Hollywood Doll Mfg. Co. The doll has painted features, mohair wig, and jointed arms and legs. She is all original and is copyright 1944. She is the Pisces doll (one of a set of twelve made to represent each Zodiac sign). A horoscope for Pisces was also included in her original box ($35).

Joy Doll Co.—Doll Craft

Many composition ethnic costumed dolls were marketed during the 1930s and 1940s. Most of these came in sizes from 9" to 13" tall. Some dolls were sold in pairs while others were sold individually. In addition to being dressed in costumes of foreign countries, many of the dolls were dressed as storybook characters or in miscellaneous costumes, which included skating, nursing, and cowgirl outfits.

It is hard to identify the makers of these dolls since the same shaped yellow tags have been found on identical dolls with either "A Doll Craft Product" or "Joy Doll Company" printed on them. The Joy firm was located in New York City and they were active in the doll making business in the 1930s and 1940s.

Regal Doll Manufacturing Co.

The Regal Doll Manufacturing Co., of New York City, was in the doll business beginning in 1919. Their factory was located in New Jersey. The company made composition baby dolls, "mama dolls," and little girl dolls. Some of their dolls carried the Kiddie Pal or Kiddie Joy trademarks. The company produced dolls with both painted eyes and sleep eyes but most dolls had molded hair.

In 1931, the company advertised two all composition dolls to compete with Effanbee's Patsy line. The dolls were called Maizie (15") and Judy (20").

The firm also produced a Shirley Temple type all composition doll in the 1930s, which was tagged "Kiddie Pal Dolly." A baby doll from this same period was tagged "Kiddie Pal Baby."

15" Regal Kiddie Pal Baby, circa 1930. She has a composition head, sleep eyes, closed mouth, molded hair, curved full composition arms and legs, and a cloth body. She is all original with her tag. She wears an organdy dress and bonnet, cotton slip, underpants, and booties. Her tag reads "Love Me/Kiddie Pal Baby." She also still has her original price tag of $1.00 from the Mabley and Carew Co. of Cincinnati, Ohio ($200).

Left:
9" composition dolls made by Joy Doll Co. and Doll Craft, circa late 1930s-early 1940s. The middle doll has been seen with the same yellow tag as the other two dolls but it read "Scarlett/A Doll Craft Product." The basic jointed doll has painted eyes, closed mouth, and a mohair wig. The tagged Joy doll on the right represents Norway and the tagged Joy doll on the left is "The Girl of the Golden West" ($50-75 each). *From the collection of Marge Meisinger. Photograph by Carol Stover.*

20" Regal "The Judy Girl," circa 1930. She has tin sleep eyes, closed mouth, molded hair, composition head and shoulder plate, arms and legs, and a cloth body. She is marked "The Judy Girl" on the back of her shoulder plate. This was one of many Patsy look-a-like dolls made in the 1930s. She has been redressed ($150+). *From the collection of Marge Meisinger. Photograph by Carol Stover.*

Reliable Toy Company Limited

The Reliable Toy Company, of Toronto, Canada, was founded in 1920 as the Canadian Statuary and Novelty Company. When the firm first went into the doll business, they imported bisque heads from Germany. In the 1920s, they began producing their own composition dolls. Beginning in the 1930s, Reliable had an agreement with the Ideal Company to enable them to use some of the American company's molds to make Reliable dolls. These included the famous Ideal Shirley Temple (see Personality Dolls chapter).

The Reliable dolls came in all price ranges and although some of their dolls were quite cheaply made, others—like the Barbara Ann Scott ice skating dolls—were of high quality. The company's dolls included the Royal Mountie, World War II Nurse (see Military Dolls chapter), Aviator, Her Highness, Indians, Eskimos, Wettums, a variety of baby dolls, dolls dressed in foreign costumes, and Patsy type dolls. Most of the dolls are clearly marked so they are easy to identify.

Reliable made fine hard plastic dolls, as well as plastic dollhouse furniture, in the 1950s.

12-1/2" Reliable composition Indian, circa late 1930s-early 1940s. He has painted eyes, closed mouth, black mohair wig, and is jointed at shoulders and hips only. Marked on the back of his head is "RELIABLE/MADE IN/CANADA." He is all original unless he once had shoes ($100-125).

15" composition Reliable "Her Highness" doll, circa late 1940s. She has sleep eyes, open mouth with teeth, a brown wig, and is fully jointed. She appears to be all original including her cape. Her red banner reads "Her Highness Coronation Doll." She is marked "Reliable" on the back of her neck ($250+).

Left:
18" Canadian "Mountie" made by Reliable, circa early 1940s. He has painted eyes, closed mouth, molded hair, composition flange head and upper arms with a cloth body and legs. He is marked on the back of the neck "RELIABLE/MADE IN/CANADA." The hat, belt, and sash are replacements, otherwise his costume is original. The doll represents a member of the Royal Canadian Mounted Police ($150-200).

Roberta Doll Co. Inc.

The Roberta Doll Co., located in New York, was active in the doll market during the 1940s and 1950s. Most of the dolls are not marked, so only the tags and boxes identify the dolls. The firm's later dolls of hard plastic seem easier to find.

It is known that all composition little girl dolls were made in the late 1940s. It is likely that baby dolls of composition were also produced.

The company's most sought after doll is the hard plastic Lu Ann Simms doll, circa 1953.

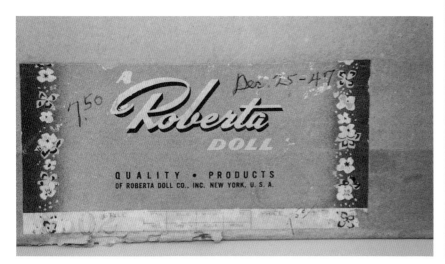

The Roberta Doll Co. box that came with the bride doll offers the following information: "A Roberta DOLL/QUALITY PRODUCTS/OF ROBERTA DOLL CO. INC. NEW YORK, U.S.A." The original owner has added the date it was received as a Christmas gift on "Dec. 25 - 47" and the 7.50 price still remains on the box.

14" composition Roberta bride doll, circa 1947. She has sleep eyes, closed mouth, mohair wig, and is fully jointed. She is not marked but comes with her original box. She is all original wearing her bride dress, long slip, underpants, shoes, socks, veil, and carrying her bouquet ($150-185 with box).

Sun Rubber Co.

The Sun Rubber Co., of Barberton, Ohio, was very active in doll production in the late 1930s, 1940s, and 1950s. The company produced rubber dolls of various kinds. Although most of these dolls were manufactured of soft rubber, some had heads of hard rubber with soft rubber bodies. Both designs were sold as "drink & wet" dolls.

The Montgomery Ward catalog for 1937 offered the 11" all soft rubber dolls for only 89 cents complete with a small layette. The 11" rubber doll, with a hard rubber head, sleep eyes, and a layette, was priced at $2.79.

The most famous of the Sun Rubber dolls from the era was the 1949 Amosandra doll (see Personality Dolls chapter). This doll was based on the baby from the Amos and Andy radio show. The firm had another big winner in the 1950s with the rubber Gerber baby doll, which was based on advertising from Gerber Baby Food of Fremont, Michigan.

11" Sun Rubber "Drink & Wet" doll, circa late 1930s. The rubber body has completely collapsed and the original clothing and head have been placed on a more modern vinyl baby body. The doll has sleep eyes, an open nursing mouth, and molded hair. The back of the rubber body is marked "MFD BY/THE SUN RUBBER CO./ BARBERTON O. U.S.A./PAT 2118682/PAT 2160739." The rear of the rubber body contains a hole so the water could drain out of the doll. The clothes consist of an organdy dress and slip, and a diaper made like the real baby diapers of the era ($25 in this condition).

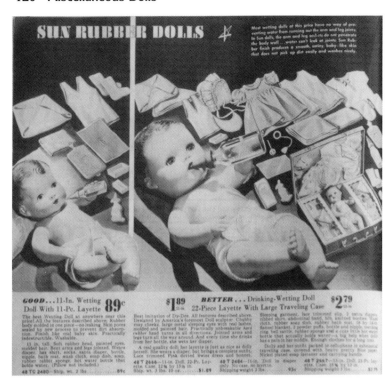

The 1939 Montgomery Ward catalog featured two different styles of Sun Rubber "Drink & Wet" dolls. The cheapest doll was 11" tall and was made of all rubber with painted eyes. It sold for 89 cents and included a layette. The most expensive doll was priced at $2.79 and included a hard rubber head with sleep eyes and a nicer layette. *From the collection of Marge Meisinger.*

Three-In-One Doll Corporation

The Three-In-One Doll corporation, located in New York City, produced the three-faced Trudy doll in 1946. The doll had a composition head with three faces. One side was "smily," one was "weepy," and one was "sleepy." Most of the dolls were made with cloth bodies and limbs. Sears advertised the 14" doll in 1946 for $5.39. The doll was also produced with composition limbs in both a 14" and a 20" size.

The original tag read "An Elsie Gilbert Creation. Patents Pending, TRUDY."

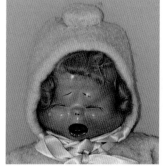

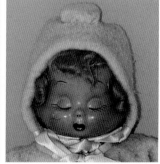

Left:
The Three-In-One Trudy doll was advertised in the Sears Christmas catalog for 1949. The doll had three faces and sold for $4.69. She had a composition head, composition arms and legs, and a cloth body.

Right:
15" Three-In-One Trudy doll, circa 1946. This model featured a composition head with three faces, and the rest of the doll was cloth. A knob on the top of the doll's head could be turned to show the different doll faces. They included sleepy, weepy, and smily expressions. The doll's features were painted. She has been redressed in a copy of her original clothing ($125+).

Uneeda Doll Co.

The Uneeda Doll Co. of New York City was established in 1917. The firm made a variety of inexpensive composition dolls during the 1930s and 1940s. Their most collectible doll of the era was the Rita Hayworth Carmen doll from 1948 (see Personality Dolls chapter).

The firm marketed a line of composition little girl, baby, and toddler dolls. The 12" toddlers were offered as a set of boy and girl twins and the same dolls, in different costumes, were sold individually.

One of the firm's most unusual baby dolls was marketed in 1936. Under the name "Sweetums," the doll was made in five sizes: 11", 13", 15", 17", and 19". The doll was a "drink & wet" doll but it was not made of rubber. Instead, the doll had a composition head, arms and legs, and a soft body. It is likely that few of these dolls survived after their "child mothers" accidentally spilled the "milk" from the bottle all over the composition head and limbs.

The Uneeda dolls were reasonably priced so they remained in business for many decades. The company had a big hit in the late 1950s when they marketed the fully jointed Dollikin dolls.

12" composition twin toddlers by Uneeda, circa 1940s. The dolls have sleep eyes, closed mouths, molded hair, and are fully jointed. They are all original with their box and tags. The box is labeled "Beauty-Quality A Uneeda Doll Est. 1917, Uneeda Doll Co. Inc. New York City, Made in U.S.A." The tags carry most of that same information. They are dressed in red and white outfits which include matching hats and shoes. The dolls sold for $3.98 a pair in 1949 ($400+). *Courtesy of Frashers' Doll Auctions, Inc.*

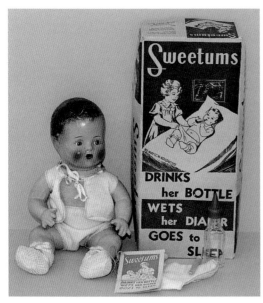

12" Uneeda Sweetums "drink and wet" doll from 1936. The doll has a composition head, full arms and legs, and a cloth body. She has sleep eyes, an open mouth, and molded hair. The original box lists the patent as 2,040.201 from May 12, 1936. The cloth body, which was especially made to resist water, is unusual for a drink and wet doll. The composition head, arms, and legs were also prone to water damage. It is likely that not many of these dolls survived the bottle feedings. Sweetums is dressed in her original diaper and shirt and has her original bottle. Her tag reads "Sweetums/DRINKS her BOTTLE/WETS her DIAPER/GOES to SLEEP." The patent date is also included on her tag ($150-200 in this mint condition).

Well Made Doll Co.

The Well Made Doll Co., located in New York City, was established in 1919. The firm made cheaper lines of dolls. One of their products, from circa 1940, was an all composition "Q-T Drinking and Wetting Baby Doll." Although the mouth contained a metal plug that was supposed to keep liquid from coming in contact with the composition head, it was not a successful deterrent. A hole in the buttocks allowed the water to run into the diaper, which then spread liquid all over the doll's bottom. It is hard to imagine a manufacturer deciding to make a "drink & wet" doll of all composition but the doll was very inexpensive so that was probably the advantage of such a product.

Right:
11" Q-T Drinking and Wetting composition Baby Doll made by the Well Made Doll Co., circa late 1930s - early 1940s. She has painted eyes, open nursing mouth, and molded hair. Her mouth has a metal rim to prevent the liquid from damaging the composition. She was packaged in a cardboard suitcase along with underwear, dress, bonnet, socks, diaper, teething ring, rattle, hot water bottle, sponge, and bottle ($100-135).

Unidentified Dolls

Since so many companies manufactured dolls during the 1930s and 1940s, it is very difficult to identify the makers of many dolls unless the dolls are marked or well known. To complicate matters further, many firms purchased basic dolls from other manufacturers and then finished the dolls themselves. In some cases, this meant the company added only clothing to the basic doll; in other cases, the painting, stringing, and wigs were all supplied by firms who did not actually manufacture the dolls.

18" unmarked composition doll in her original clothing. She has sleep eyes, closed mouth, and a mohair wig. She is wearing a red satin and black lace dress with a matching black lace head piece ($150+).

9" composition boy and girl dolls dressed in native costumes, circa 1940s. They have painted eyes, closed mouths, and the girl has a mohair wig while the boy's hair is molded. They are jointed at the shoulders and hips and are wearing their original clothing, shoes and socks. Neither doll is marked (pair $80-110).

The Montgomery Ward catalog of 1940 advertised an ethnic composition "Kinky" doll typical of the era. She sold for only 23 cents and featured three black yarn pigtails. *From the collection of Marge Meisinger.*

9-1/2" composition Dixie ethnic doll, circa 1940. This doll looks very much like the doll pictured in the Montgomery Ward 1940 catalog. She has painted eyes, closed mouth, and molded hair along with three tuffs of mohair on her head. She is jointed at the shoulders and hips. She is all original with her tag, which reads "Dixie/Made in U.S.A." ($65+).

Character Dolls

Many dolls marketed in the 1930s and 1940s can be considered character dolls. Some were based on comic characters featured in the popular newspaper comic strips; others were based on artist designed characters used in advertising, print, and other types of media. Another category of character dolls from the era were those based on the popular animated figures from Walt Disney's motion pictures. Other dolls placed in this chapter include unusual dolls that were designed by doll artists and were not based on anything but the artists' creative imaginations.

Advertising Dolls

Campbell Soup (Campbell Kids)

The Campbell Kids dolls were based on drawings by Grace Drayton that were used to advertise Campbell Soup. The first advertisements were circa 1905. It wasn't long before E.I. Horsman Co. began producing Campbell Kid dolls. These early dolls had composition heads and hands, with cloth bodies and legs. They were marked "E.I.H. © 1910" on the back of their heads.

By the end of the 1920s, the Campbell Kid dolls were being marketed by the American Character Doll Co. These dolls had composition heads, full composition arms and legs, and cloth bodies. They were 12" tall and were marked on the neck "Campbell Kid." The same company also produced an all composition Campbell Kid. All these dolls have molded hair and painted features.

The Horsman firm again began making composition Campbell Kids in 1948. These dolls were 12" tall, fully jointed, with molded shoes and socks, painted features, and molded hair. These later dolls were unmarked.

12" Horsman composition Campbell Kid dolls, circa 1948. The dolls have painted features, closed mouths, molded hair, molded shoes and socks, and are fully jointed. They are all original in mint condition and include their original tags. The tags read "Campbell's KID/A HORSMAN DOLL/Permission of Campbell Soup Company." A can of Campbell's tomato soup is pictured on the tag. The dolls are unmarked ($350+ each). *Dolls from the collection of Lois Jakubowski. Photograph by Robert Jakubowski.*

Lee (H.D) Company (Buddy Lee)

The Buddy Lee dolls that were produced to promote products for the H.D. Lee Company have been some of the most popular advertising dolls ever made. Although the dolls were never used as premium dolls, they were sold by the Lee company to advertise their clothing. The first dolls were composition dolls sent to dealers to place in store windows. When the dealers were finished using the dolls, they were sold to customers. These dolls proved so popular the company began selling them as part of their regular line of products. In 1949, the dolls were re-designed and were then made of hard plastic. These new dolls were 13" tall. In addition to cowboy and farm outfits, several different uniforms were made for the dolls. Eventually, the hats and clothing were contracted out to other companies.

13" composition Buddy Lee made to advertise the H.D. Lee company, circa 1940s. He has painted features, molded hair, molded shoes, and jointed arms only. He is wearing "Lee" jeans and a denim shirt. His clothes are labeled "Buddy Lee Union Made" ($300+).

Comic and Cartoon Dolls

Betty Boop

The cartoon character Betty Boop was originated by animator Max Fleisher for the Fleischer Studios circa 1930-31. She soon became a very popular movie character for Paramount Pictures and many different cartoons were produced.

With the popularity of the movies, tie-in products in Betty Boop's image were soon being developed and marketed. One of the most successful of these enterprises was the Betty Boop doll first made in 1932. She was designed by Joseph Kallus and sold by his Cameo Doll Co. of Port Allegany, Pennsylvania. The doll was made of composition and wood and stood 12" tall. The arms, legs, and lower torso were wood while the rest of the doll was composition. The doll was jointed at the neck, shoulders, hips, and knees. The Betty Boop doll had molded hair and painted features. Her clothes were also painted on the body. The dolls sold for $1.00 to $1.49 each.

The last Betty Boop cartoon was produced in 1939 and interest in the figure died until it was revived in the 1970s with new tie-in merchandise. Products and new dolls continue to be marketed in Betty Boop's image but the early dolls of the 1930s remain as the most collectible of the Betty Boop merchandise.

Blondie

The "Blondie" comic strip was begun on September 15, 1930 by Murat (Chic) Young for King Syndicate. In the story line, Blondie married Dagwood Bumstead in 1933 and they had two children. Alexander was born in 1934 and Cookie joined the family in 1941. Daisy, the dog, completed the household.

Several dolls were marketed in the 1930s and 1940s based on these famous characters. All of the dolls are rare. The Knickerbocker Toy Co. produced a composition 11" Blondie doll with a mohair wig and painted eyes, as well as a 14" Dagwood and a 9" Alexander. Both the male dolls were marked "Knickerbocker Toy" on the back of their heads. The clothes on all the dolls were removable but the shoes and socks were molded on the Alexander and Dagwood dolls. These two dolls had painted features and molded hair.

Another set of Blondie related dolls, made by Columbia Toy Products, appeared on the market in 1947. An 18" Dagwood doll and a smaller Alexander doll were included in a series of comic cloth dolls produced by the firm. Both dolls had molded buckram mask faces, painted features, fur-like hair, and tufts of straw-type hair sticking up from their heads.

Joe Palooka

"Joe Palooka" was a comic strip developed by Ham Fisher in 1928. Joe was a boxer with a manager named Knobby Walsh. The *New York Mirror* syndicated the feature in 1931. Later, both a radio program and a television series about Joe Palooka were produced.

Several dolls were made during the 1950s to represent characters from the Palooka strip (see *Dolls and Accessories of the 1950s*). Only one doll was produced in the 1940s. It was the Humphrey Doll called "The Happy Blacksmith, Joe Palooka's Pal." The doll had a plastic type mask face with a cloth body and was circa late 1940s. His face appeared to be made of the same material as Ideal's "Talkytot" and may have been produced by that company.

Little Lulu

"Little Lulu" had her beginning in June 1935 as a single panel cartoon in the *Saturday Evening Post*. The comic was drawn by Marge (Marjorie Henderson Buell). The "Little Lulu" character eventually was featured in comic books in 1945 and in a newspaper strip from 1955-67. Other characters from the strip include Tubby and Alvin.

Little Lulu cloth dolls were produced by the Knickerbocker Toy Co. circa late 1930s and early 1940s. Their original tags read "Little Lulu/From/The Saturday Evening Post" on one side and "Knickerbocker Toy Co. Inc./ New York" on the other. The dolls came in at least two sizes, 13" and 17". They were dressed in several different dress styles.

11" composition "Blondie" by Knickerbocker Toy Co., circa late 1930s. She has painted eyes, closed mouth, a mohair wig, and is fully jointed. She is all original and comes with her original tag and box. The tag reads "BLONDIE/BABY DUMPLING/DAGWOOD/AND/DAISY." The company also marketed all composition dolls in the images of Dagwood and Baby Dumpling (apparently Alexander). The tag indicates that a Daisy dog was also made. The printing on the box reads "Knickerbocker Toy Co. Stuffed Animals, 606/2 Blondie." Since this doll is not marked, she is very hard to identify without her tag or box (sold at auction for $1,680). *Courtesy of Frashers' Doll Auctions.*

15" cloth Little Lulu made by Georgene Novelties, Inc. She has a buckram type molded face, painted features, and thread hair. She is all original complete with her plastic purse. The purse is marked "Little Lulu by Marge/19©44/M-H.B." Although the 1944 copyright appears on the purse, the doll was actually purchased from a retail store in 1954. The matching 13" Alvin was also made by Georgene Novelties and carries a copyright of 1951 on his original tag. He is wearing his original clothing, including a beany hat. Both dolls have mitten hands and shoes made as part of their feet. The dolls were based on the characters originated by Marjorie H. Buell (both dolls with some wear Lulu $100+, Alvin $200+).

The Little Lulu dolls made by Georgene Novelties Inc. were made later, circa late 1940s or early 1950s. Tubby and Alvin companion dolls were also marketed by Georgene Novelties but the original tag on Alvin dates him from 1951. These dolls vary in size from 12-1/2" (Alvin) to 14-1/2" (Lulu and Tubby). The Lulu dolls in Hawaiian and Cowgirl outfits are the hardest to find.

Little Orphan Annie

Harold Gray began the "Little Orphan Annie" comic strip on August 5, 1924 for the Chicago Tribune-New York News Syndicate. Besides Annie, leading characters in the strip were Daddy Warbucks, Punjab, and Annie's dog Sandy. "Little Orphan Annie" was so popular that a radio show, movies, and a Broadway musical were all produced based on this comic strip.

Many "Little Orphan Annie" dolls have been made through the years, particularly during the late 1970s and early 1980s when the Broadway show and the movie musical added to Annie's popularity. At least two different dolls were produced during the 1930s. These include a composition shoulder head Annie with composition arms, painted features, and molded hair from 1931. The doll had a cloth body and legs. She was advertised in the Sears catalog in 1931 in a 14-1/2" size for 45 cents. Another all composition Annie was marketed in 1936. This doll came in 10" and 12" sizes. She had painted features, molded hair, and was fully jointed. Although the dolls were unmarked, they have been attributed to the Ralph A. Freundlich firm. A Sandy dog came with the doll.

12" composition "Little Orphan Annie," from the comic strip of the same name, circa late 1930s. Made by Freundlich, the doll has painted eyes, closed mouth, molded hair, and is fully jointed. She is all original but was originally sold with her dog Sandy. She is pictured with a "Little Orphan Annie" metal stove made by Marx circa 1940 (doll $175+, stove $75+).

Popeye

The Popeye character began as part of the "Thimble Theater" comic strip, which was started by Elzie Crisler Segar for the W.R. Hearst chain in 1919. Popeye became a part of the strip in 1929. Other important characters were Olive Oyl, Jeep, Swee'pea and Wimpy. Paramount took advantage of the characters' popularity by making animated movie cartoons beginning in 1932.

Due to the success of the Popeye character, several different Popeye dolls were marketed during the 1930s and 1940s. One of the first was an all cloth Popeye doll produced by the Effanbee Doll Co. in the early 1930s. This doll was 16" tall and was dressed in the traditional Popeye costume.

Several different wood segmented Popeye dolls were also produced in the 1930s. The earliest bears a copyright marking of 1932 by King Features Syndicate. The doll was 10" tall and featured a jointed wood-composition body. The clothes were painted on the doll. It was made by J. Chein Company of Harrison, New Jersey. The Cameo Doll Co. marketed a similar 14" doll called POP/EYE in 1936. It had a composition head and body with wood segmented limbs. In 1933, the Cameo firm produced a Jeep doll made of composition and wood. The 13" doll was fully jointed and was also based on a character from the "Popeye" strip.

Skippy

"Skippy" was first drawn by Percy Crosby for the original *Life* magazine in 1919. Because people seemed to enjoy the character, Crosby began a "Skippy" syndication strip in the early 1920s. King Features took over the distribution in 1928. Skippy was a ten-year-old boy who always wore a plaid hat and short pants. In 1931, Jackie Cooper starred in a successful Skippy movie based on the comic character.

The Effanbee Doll Co. produced several models of Skippy dolls beginning in 1929. Most of the dolls were all composition but some had cloth bodies and upper legs. All of the dolls had painted eyes and molded hair. Effanbee issued many different Skippy dolls through the years. He was still being sold in 1943 dressed in military uniforms instead of his "little boy" clothing.

14" composition Effanbee Skippy, circa early 1930s. He has painted eyes, closed mouth, molded hair, and is fully jointed. He is all original with his pin. He is marked "EFFANBEE/SKIPPY/© PL CROSBY" on his neck and EFFanBEE/PATSY/Pat. Pending/DOLL" on his body. His pinback reads "EFFanBEE DOLLS/I AM/SKIPPY/TRADE MARK/ THE REAL AMERICAN BOY." He is pictured with a "Big Little" Skippy book by Percy Crosby, published by Whitman Publishing Co. in 1930 (doll $450-500, book $20). *From the collection of Jan Hershey.*

Superman

"Superman" was the first costumed "super hero." He was developed by Jerry Siegel and Joe Shuster (both seventeen years old at the time) as a comic book hero in 1938. Superman came from another planet and had a dual identity. He was Clark Kent, a newspaper reporter, in everyday life and became Superman in times of trouble. The character was featured on a radio show, in the movies, and in a television series.

A Superman doll designed by Joseph Kallus was marketed by Ideal in the early 1940s. The doll was 13" tall, had a composition head and upper torso, and a wood ball-jointed body. Superman's familiar red and blue outfit was painted on the body. He also came with a cloth cape.

Another Superman composition doll was issued in the 1940s as well. The doll looked like a regular toddler doll with molded hair but was dressed in a Superman costume. The company who made the doll is unknown.

Winnie Winkle (Denny Dimwit)

Created by Martin Branner, the "Winnie Winkle" comic began in a daily strip on September 20, 1920. The *Chicago Tribune-New York News* carried the syndication. The story line followed the adventures of Winnie Winkle, a young working woman at a time when most women were not employed. Other characters included her brother Perry, Ma and Rip Winkle, Denny Dimwit, and Will Wright.

An unusual doll was made circa 1948 in the image of Denny Dimwit. The 11" composition doll had molded hair, hat, clothes, and painted features. The head and body nodded and wobbled when the doll was touched or moved. The original box identified the character as Denny Dimwit. A similar doll was also produced to honor Sammy Kaye, whose band motto was "Swing and Sway With Sammy Kaye." The doll's name was Bobbi Mae. Both dolls were produced by the Wondercraft Company of New York.

Disney Character Dolls

Snow White

The success of Walt Disney's animated film, *Snow White and the Seven Dwarfs*, completed in 1937, had a big impact on the doll world. With the picture's general release in 1938, many products were produced to tie in to this popular movie. Dolls, coloring books, paper dolls, and puppets were just some of the many toys made to take advantage of the public's interest in anything to do with the Disney production. The Ideal Company obtained the rights to produce Snow White and the Seven Dwarfs dolls based on the Disney characters. Several different Snow White dolls were made, including a model with molded hair and painted features, a doll with a black mohair wig and sleep eyes, and an all cloth version. The molded hair dolls came in sizes of 14", 17-1/2", and 19". The wigged doll was issued in sizes of 11-1/2", 13", 14", 18", 22", and 27". The cloth dolls were 16" tall.

The Ideal Dwarfs were 12" tall and had oilcloth type hands and faces, with the rest of the doll made of stuffed cloth. These dolls were made in the images of all the Disney dwarfs: Doc, Happy, Sneezy, Sleepy, Bashful, Grumpy, and Dopey. The 1938 Sears catalog also pictures an 11-1/2" Dopey doll made with a composition head and stuffed body. He sold for 95 cents. A larger 20" Dopey ventriloquist doll was sold that same year. He had a composition head and hands, with a cloth body, arms, and legs.

Knickerbocker Toy Co. was also authorized to make Disney Snow White and Seven Dwarfs dolls. The Snow White doll is marked "Walt Disney 1937, Knickerbocker Toy Co. New York." The composition Snow White was 15-1/2" tall with painted features, molded hair, and ribbon. The 9" dwarfs were also composition with painted features and applied beards.

The Madame Alexander Co. was another firm that sold authorized Disney Snow White products in 1938. Included in the Alexander items was a series of 12" marionettes featuring Snow White herself. She had a composition head, arms and legs, and a cloth body. The Snow White doll issued by Alexander in 1938 used the Princess Elizabeth face mold, with the doll dressed in a Snow White costume. The composition doll came in 13" and 16" sizes and had a black human hair wig. Although she was marked "Princess Elizabeth" on the back of her head, she originally came with a paper tag identifying her as "Walt Disney's Snow White."

Most of the other doll companies jumped on the bandwagon and produced Snow White dolls of their own. Because the Snow White character was based on a fairy tale, Disney could not prevent others from making generic dolls, as long as the firms did not try to copy any of the Disney designs.

Effanbee marketed an open mouth American Child doll dressed as Snow White in the late 1930s. Since the firm was not authorized by Disney to make a Snow White related to the film, Effanbee labeled their doll as Snow White from Grimm's Fairy Tales.

Horsman offered a 19" composition girl doll dressed in a Snow White looking costume for sale in the late 1930s. She had a black wig, sleep eyes, and wore a long dress and cape. The bodice of the dress was red velvet.

Other lesser known firms also marketed composition dolls with molded hair and hair ribbons that had the look of Snow White.

The Sears 1938 Christmas catalog featured the Ideal Snow White and Dopey dolls from the Disney *Snow White and the Seven Dwarfs* new motion picture. The composition Snow White was offered in two sizes of 11-1/2" and 13", priced at $1.98 and $2.98. The Dopey doll was 11-1/2" tall and had a composition head and a stuffed body. He sold for 95 cents.

Character Dolls 127

On the left is an 18" composition Ideal "Snow White" from 1938, made using the Shirley Temple body. She has sleep eyes, open mouth with teeth, and mohair wig. She wears a taffeta gown with velvet bodice and cape. The doll on the right is a 16" Ideal cloth "Snow White" with an oilcloth type mask face and arms. Her features are painted and she has a human hair wig. She is wearing an organdy skirt with a cotton bodice. Both dolls have their names and drawings of the seven dwarfs on the bottom of their skirts. A composition doll wearing a similar costume was also made with hair and a ribbon molded on her head (composition $400+ with crazing, cloth $300+). *From the collection of Marge Meisinger. Photograph by Carol Stover.*

Knickerbocker set of composition "Snow White and the Seven Dwarfs" (Doc not shown). 15" Snow White has painted eyes, closed mouth, molded hair, molded ribbon in hair, and is fully jointed. She wears her original costume including her cape. She is marked "Walt Disney 1937, Knickerbocker Toy Co. New York." The 9" dwarfs are all composition, jointed at the shoulders. They have painted features, applied beards, molded shoes, and are all original. Four have their original boxes and tags. The tags read "Walt Disney's Snow White and the Seven Dwarfs." Two different boxes were used for the dwarfs. Both are marked Knickerbocker Toy Co. with the name of the individual dwarf stamped on the box (sold at auction for $2,730). *Courtesy of Frashers' Doll Auctions, Inc.*

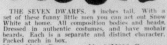

The N. Shure Co. catalog of 1938 carried an advertisement for what may be the Knickerbocker Snow White and Seven Dwarfs. The composition Snow White was 15" tall and the copy states that the doll has glass eyes and a wig. The Knickerbocker 9" dwarfs were all composition with mohair beards. The Dopey doll pictured is Ideal's 20" ventriloquist doll. He had a composition head and hands with a cloth body, arms, and legs. His mouth is hinged and could be operated by a drawstring. *From the collection of Marge Meisinger.*

Character Dolls

Alexander authorized 16" composition Snow White, circa late 1930s. She has sleep eyes, closed mouth, black wig, and is fully jointed. She is mint with her original Disney wrist tag and label on her box. The tag reads "Walt Disney's Snow White Dolls by Madame Alexander." The printing on the box states "Walt Disney/Snow White/All Rights Reserved/Madame Alexander." The original price of $4.98 remains on the box (MIB $850+). *Doll and photograph from the collection of Veronica Jochens.*

Pinocchio

Walt Disney continued the success of his animated films with the release of *Pinocchio* in 1940. The plot involved a puppet who wanted to become a real boy. Some Disney fans say that *Pinocchio* is Walt Disney's greatest animated film.

The movie, like all Disney features, inspired many tie-in products. Current collectors are especially interested in the dolls. Several different companies produced a variety of Pinocchio dolls. The dolls made in the late 1930s and early 1940s include those from Crown Toy Co., Ideal, and Knickerbocker. Ideal's Pinocchio dolls ranged in size from 8" to 20". Their dolls had composition heads, bodies, and gloved hands along with wood segmented limbs. The Ideal firm also manufactured a Jiminy Cricket doll in an 8" size. This doll was constructed of wood. All of these dolls were licensed by Walt Disney.

The Knickerbocker Pinocchio was also sold as a Walt Disney product circa 1939. The dolls were marked on the head "Pinocchio/Knickerbocker Toy Co./Walt Disney Toy Product." These dolls were made of composition and were fully jointed. The hat and clothing of this doll were removable.

The Crown Toy composition Pinocchio dolls came in several styles. One model had a painted hat and clothing and was jointed only at the arms. Another doll featured removable cloth clothing and was jointed at both the shoulders and hips. These dolls were from 8-1/2" to 9-1/2" tall. Both models were marked "W. DISNEY PROD." as well as with "CROWN TOY CO." identification.

9" composition Pinocchio made by the Crown Toy Co., circa 1939-40. He has painted eyes, closed mouth, and molded hair. He is jointed at the shoulders and hips and has molded shoes and gloves. He is missing his hat and his original shirt. On the back of the head, he is marked "PINOCCHIO/W.D.PROD./MADE IN U.S.A./CROWN TOY CO." Shown with book: *Walt Disney's Pinocchio*, Golden Press. © Walt Disney Prod. ($150+ not original).

20" Ideal Pinocchio doll, circa 1939. He has a composition head and a wooden segmented body with painted features. His clothes are molded on his body except for a felt cap and bow tie. He is marked on the front "Pinocchio/Des. & © by Walt Disney/Made by Ideal Novelty & Toy Co." Marked on his head is "© P. WDP/Ideal Doll/Made in USA." ($800+). *Doll from the collection of Marge Meisinger. Photograph by Carol Stover.*

Miscellaneous Character Dolls

Joseph L. Kallus Dolls-Cameo Doll Co.

The Cameo Doll Co. was founded by Joseph L. Kallus in 1922. The firm was very active in producing and selling Rose O'Neill designed dolls, including Kewpies, Scootles, and Giggles. The company plant was located in Port Allegany, Pennsylvania from 1932 until 1970. The original plant burned in 1935 and another plant was built.

Besides Rose O'Neill designs, the company marketed many other dolls including those designed by Joseph Kallus. These composition dolls included Betty Boop, Pinkie, Champ, Margie, Joy, and POP/EYE. Pinkie was sculptured by Kallus in 1930. The 10" doll had a composition head, wooden body, and segmented limbs. She had painted features and molded hair. Margie, a similar doll, was issued about the same time. Margie came in 6", 10", and 15" sizes and was advertised in *Playthings* magazine in 1930. The larger dolls were made of composition while the smaller models had composition heads and wood segmented bodies and limbs. The Kallus designed dolls all had painted features and molded hair. Joy, another Kallus design, was marketed in 1933. She was similar to the other dolls but she had a ribbon loop molded on her composition head. She came in 11" and 15" sizes. The larger size doll had a head made of composition as were her body, hands, and lower legs. The doll's limbs were wood segmented. The smaller doll had a composition head and hands with a wood body and segmented limbs.

Another winner for the firm was Miss Peep, sculptured by Kallus and manufactured in vinyl in the 1960s.

In 1970 the Cameo Doll Co. was sold to the Strombecker Corp. of Chicago, Illinois.

Rose O'Neill Dolls

The Kewpie character was designed by Rose O'Neill in 1909 to be used in drawings for the *Woman's Home Companion* magazine. Joseph Kallus helped O'Neill translate her drawings into dolls. The first dolls, made of bisque, were manufactured in Germany in 1913. By the 1930s and 1940s, Kewpie dolls were being manufactured in both cloth and composition. In the 1930s and early 1940s, composition Kewpie dolls were produced by the Cameo Doll Co. By 1933, these dolls had jointed necks, arms, and legs.

In the Sears catalog in 1930, cloth Kewpies were being advertised in 11-1/2" sizes. These dolls, made by Krueger, sold for $1.25 each (see Cloth Dolls chapter).

In the later 1940s, the Effanbee firm secured the rights to produce composition Kewpie and Scootles dolls. Through the years, some composition Kewpies and Scootles dolls were made in brown models as well as flesh color.

The Rose O'Neill designed Scootles dolls were first made of bisque in Germany in 1925. Joseph Kallas also helped with this process. His Cameo Doll Co. began making the dolls of composition in 1934 and they were advertised in a *Playthings* ad in 1935. The dolls came in a variety of sizes from 7-1/2" to 20" tall.

Another Rose O'Neill designed doll was issued by the Cameo Doll Co. circa 1947. The 12" composition "Giggles" doll had molded hair, painted features, two holes in the back of her head for a hair ribbon, and was fully jointed. She was not marked. She is quite a rare doll today.

13" composition Cameo Kewpie like the one advertised in the 1946 Sears Christmas catalog. She has painted features, molded hair with a top knot, and is fully jointed. She is all original and is not marked ($300).

12-1/2" Rose O'Neill composition brown Scootles. She has painted eyes, closed mouth, molded hair, and is fully jointed. She wears what appears to be her original dress but her shoes and socks have been added. She is not marked ($500+).

Sears advertised composition Rose O'Neill Kewpie and Scootles dolls in their 1946 Christmas catalog. The Kewpie was 12-3/4" tall and sold for $2.59. Scootles was 16" tall and sold for $4.98. Both dolls had labels from the Cameo Doll Co. *From the collection of Betty Nichols.*

Character Dolls

Through the years, the Rose O'Neill character dolls have given both children and collectors lots of pleasure. The Kewpie character continues to flourish and it is on its way to becoming a true American icon.

Herman Cohen (House of Puzzy)

As the head of the House of Puzzy (located in Baltimore, Maryland), Herman Cohen marketed a pair of unusual character dolls, circa 1948, that remain very collectible today. It is not known if the dolls represented real characters or if the designer, like Joseph Kallus, produced the dolls because they were different than any then on the market. The Puzzy and Sissy composition dolls are both 15" tall with molded hair and painted features. Both dolls are fully jointed. They were marked on the back of their heads "H. of P. USA."

10" composition Kallus designed "Pinkie" on the left and 10" composition "Margie" on the right. Both dolls were made by the Cameo Doll Co. They have painted features, molded hair, and composition socket heads with wood segmented bodies. Each doll is marked with its name on the front of its body ($250-300 each). *Courtesy of Frashers' Doll Auctions, Inc.*

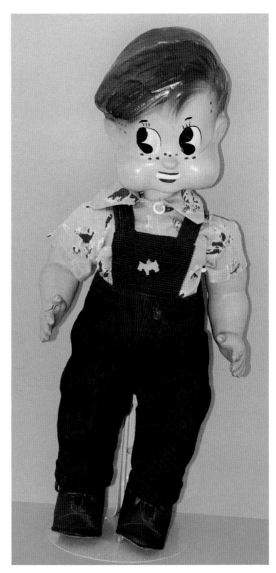

15" composition Puzzy made by the House of Puzzy. He has painted features, molded hair, and is fully jointed. He is marked "H. of P. U.S.A." on his head. He has been redressed. He originally wore short pants, white shirt, and tie ($300-350). *Doll from the collection of Jerry and Gayle Reilly.*

15" composition Sizzy (companion to Puzzy) made by the House of Puzzy. She has painted features, molded hair with a top knot, and is fully jointed. She is marked "H. of P. U.S.A." on her head. She has probably been redressed ($300-350). *Doll and photograph from the collection of Jerry and Gayle Reilly.*

Raggedy Ann and Andy

The Raggedy Ann doll was patented in 1915 by John B. Gruelle (John Barton) and the first Raggedy Ann stories appeared in 1918. The earliest dolls had brown yarn hair instead of the well-known red that tops today's Raggedy Ann. They also had painted features and, of course, were made of cloth.

Several different manufacturers produced the Raggedy Ann and Andy dolls during the 1930s and 1940s. The P. F. Volland Co. was responsible for making the dolls during the early 1930s until the company went out of business in 1934. This firm also published the Raggedy Ann books.

By 1936, Molly'es Doll Outfitters was advertising both the Raggedy Ann and Raggedy Andy dolls for sale. According to author Jan Foulke, writing in the September 1992 *Doll Reader*, the Gruelle family disputed the right of the firm to produce the dolls and a lawsuit soon followed. It was settled in favor of the Gruelle family in 1938 and Molly'es was no longer able to make the dolls.

Quite a number of the Molly'es dolls were produced. Most of the dolls' noses were outlined in black and they had either shoe button or painted eyes.

The next company to become a licensed manufacturer of the Raggedy dolls was Georgene Novelties, Inc. This New York City based firm produced the dolls from 1938 until 1963. The early models made by Georgene featured a large red nose outlined in black and black shoe button eyes sewed over a white painted circle. One of the most popular models of Raggedy dolls in the 1940s was a series of the dolls made with two faces. The dolls were asleep on one face and awake on the other. A band of hair was sewed over the top of the head which showed on both sides.

The Raggedy Ann and Andy dolls have continued to be produced by a variety of manufacturers and their popularity with consumers still remains high.

18" Molly'es cloth Raggedy Andy, circa 1936-37. The doll has painted features with a red nose outlined in black, yarn hair, and is all cloth. On his chest is written in black "Raggedy Ann & Raggedy Andy Dolls/Manufactured by Molly'es Doll Outfitters." A red heart is also printed on his chest. Andy is all original ($1,200+). *Doll from the collection of "Susie's Museum of Childhood" at Bluebird Farm in Carrollton, Ohio. Photograph by Marilyn Pittman.*

18" Raggedy Ann and Andy dolls made by Georgene Novelties, circa 1938-44. The dolls have shoe button eyes, red noses outlined in black, and yarn hair. The tag reads "Johnny Gruelle's Own 'Raggedy Ann' Doll/Georgene Novelties, Inc./New York City." Raggedy Ann has a heart that says "I Love You." She is probably missing her apron, otherwise the dolls are all original (Raggedy Ann $1,000+, Andy $1,200+). *Dolls from the collection of "Susie's Museum of Childhood" at Bluebird Farm in Carrollton, Ohio. Photograph by Marilyn Pittman.*

A display of Raggedy Ann dolls and related products from "Susie's Museum of Childhood" at Bluebird Farm in Carrollton, Ohio. *Photograph by Marilyn Pittman.*

Santa Claus

Santa Claus dolls were marketed in the 1930s and 1940s just as they are today, but on a lesser scale. Most of the Santa figures produced in those decades were made of paper, celluloid, or hard plastic and could be classified as figurines instead of dolls. But there were true Santa dolls created of cloth and/or composition during this era. Most of the dolls did not have movable limbs but a few composition ones did. At least two models of all composition Santas, with molded beards, are known to collectors. One has added oilcloth boots (see *Collector's Encyclopedia of American Composition Dolls* by Ursula R. Mertz) while the other has molded boots. Both dolls have jointed arms and necks while only the doll with added boots has jointed legs. Another composition jointed Santa, pictured here, appears to be a toddler doll made into a Santa with the addition of a Santa suit and beard.

Many more cloth Santas were made during the period. Some had buckram type mask faces while others had mask faces of a later plastic-type material. Since the Santas were unmarked, the companies who produced the toys remain unidentified.

It is hoped that continued research will eventually identify the makers of these Santa Claus dolls. In the meantime, doll collectors must compete with Christmas collectors to add old Santa dolls to their collections.

16" composition Santa, circa late 1940s. The doll has a composition baby or toddler head with sleep eyes and a closed mouth. The beard has been added. The body is cloth and the doll is all original wearing red plastic boots and a red corduroy Santa suit with white trim. He is pictured with a small hard plastic Santa on snow shoes and a Santa puzzle made by the Whitman Publishing Co. (doll $135+, puzzle $15).

19" composition Santa Claus, circa 1930s. He has painted features along with a molded beard, mustache, and hat. He is jointed at the neck and shoulders but his legs are stationary. He wears molded black boots and is not marked. He has had some repainting and has been redressed ($175+).

Left:
24" tall Santa with an early plastic mask face, circa late 1940s. He has painted features and a white mohair beard and hair. His suit is red taffeta and he wears shiny black boots and a matching belt of a material similar to oilcloth. He is unmarked ($125+).

Right:
18" cloth Santa, circa 1930s. He has a molded painted buckram face, mohair wig and beard, and is all original in his red taffeta suit. The boots are made of heavy plaster-like material. He is not jointed and is unmarked ($250+).

Cloth Dolls

Although collectors continue to purchase cloth dolls from the 1930s and 1940s manufactured by Lenci, Kathe Kruse, Chad Valley, and Norah Wellings, they seem to neglect the more reasonably priced dolls produced in the United States during the same time period. Most of these dolls were made of cloth, stuffed with cotton, with molded buckram mask faces. The hair was usually thread or yarn but some dolls had mohair wigs. The hair sometimes consisted of only bangs while the rest of the head was covered by a cap. Other dolls had full wigs styled with curls or pigtails. Some of the dolls had clothes made as part of the body while the more expensive dolls had removable clothing. Other types of material used for bodies included an oilcloth or plush fabric. The shoes were either made as part of the feet, using a different kind of material, or else they were the regular removable shoes used on other dolls. The faces were painted on the flesh colored buckram or on top of an oilcloth type mask of a darker orange color.

The most desirable of these types of cloth dolls are those produced by the Alexander Doll Co., but most of the major doll firms also marketed cloth dolls that are worthy of a place in any collection. Most of the available dolls, however, are those made by Georgene Novelties, Molly'es, Krueger, Tiny Town, or lesser known firms.

The following companies produced cloth dolls in sufficient numbers to offer the collector many choices in obtaining dolls for a current collection.

Alexander Doll Co.

Madame Alexander (Beatrice Behrman) began making cloth dolls in the 1920s. By the 1930s, the cloth dolls from the company were being produced to represent various story book characters and were very nicely costumed. Some of these dolls include the following:

LITTLE WOMEN: These cloth dolls represented Jo, Meg, Beth, and Amy, the characters from Louisa Mae Alcott's book. The dolls were 16" tall, wore mohair wigs, had pressed felt masked painted faces, and mitten type hands. They were circa early 1930s.

OLIVER TWIST: This doll was made in the image of the main character in Charles Dickens's famous *Oliver Twist* book. The doll was 16" tall, had a pressed felt masked painted face and mitten type hands. Circa 1934.

DAVID COPPERFIELD & LITTLE EMILY: These 16" dolls were also Dickens characters based on the book, *David Copperfield*. They were like the other dolls with different clothing used to make each of the Dickens characters special. Circa 1934.

TINY TIM: He was based on the Dickens character from *A Christmas Carol*. Other Dickens characters sold in doll form included LITTLE NEL, PIP, LITTLE AGNES, and LITTLE DORRIT. All of these 16" dolls were made from the same basic doll design but their clothes, features, and wigs distinguished one from another. All of the dolls wore clothes tagged with the Alexander name.

ALICE IN WONDERLAND: This doll is thought to have been the first cloth doll marketed by the Alexander firm. There were several designs and sizes of Alice. She was made in a 16" size with the look of the Dickens dolls and she was also produced with a flat cotton face, painted features, and yellow yarn hair in a 22" size.

DIONNE QUINTUPLETS: The most collectible of the early cloth Alexander dolls are the dolls made to represent the Dionne Quintuplets. These dolls came in a variety of sizes from 15" to 24" tall. They had molded felt mask faces and human hair wigs. Circa 1936.

SUSIE-Q & BOBBY-Q: These dolls were advertised in the John Plain catalog for 1939. Both dolls were 16" tall, had yarn hair, and mask faces. Bobby carried an umbrella. A smaller 12"-13" set may have also been made.

The John Plain catalog for 1939 featured two of the cloth Madame Alexander dolls, Susie-Q and Bobby-Q. Both dolls were 16" tall with molded mask faces and yarn hair. Bobby carried an umbrella and a school book while Susie carried a purse. Both dolls had removable clothing. *From the collection of Marge Meisinger.*

134 Cloth Dolls

12" Alexander cloth Bunny Beau, circa 1939. A companion Bunny Belle was also marketed at the same time. The doll is missing his ears as he was made as an Easter toy and originally was a rabbit. After Easter, the owner evidently wanted a more realistic boy doll and removed his ears. He is also missing his hat; the other clothing is original. He has a molded buckram face with painted features, yellow yarn hair, mitten hands, and removable clothing ($100-150 in this condition).

18" cloth Little Shaver made by Madame Alexander, circa 1942-43. She has a pressed mask face, painted features, yarn hair, mitten hands, and a jointed neck. Wire was inserted into her limbs to make her posable. She is original except she is missing her purse ($300 with wear).

The Sears Christmas catalog circa 1943 advertised the cloth Alexander Little Shaver doll in two sizes, 11" and 15". The dolls sold for $4.38 and $6.79 each. The advertising copy described the dolls as having pressed mask faces, painted eyes, yarn hair, and flexible cotton bodies that twisted into many poses.

LITTLE SHAVER: This doll was based on Victorian children in a painting by Eloise Shaver. The dolls had wires in the arms and legs to make them posable. They were advertised in the Sears Christmas catalog circa 1943 in 11" and 15" sizes. They sold for $4.38 or $6.79 each. The ad described the dolls as having soft flexible cotton bodies, pressed mask faces, painted eyes, and yarn hair.

BUNNY BEAU and BUNNY BELLE: These dolls were made for Easter in 1939. They were 13" tall, had pressed mask faces, painted features, yarn hair, and mitten hands.

SO-LITE BABY: This was a 17" baby doll with a soft stuffed stockenette body and legs, neck jointed, inset lashes, and hair in front. A black baby was also produced. Circa early 1940s.

TEENY TWINKLE: This doll was a baby made of an oilcloth material with celluloid disc eyes.

BABBIE: This doll was 16" tall and made to represent a character from the book *Little Minister* by James M. Barrie. Katharine Hepburn starred in a film based on the book in 1934 and it is likely the doll was made at that time.

DOTTIE DUM BUNNIE: A dressed 16" rabbit, circa 1938.

TIPPIE TOE: This was an 18" cloth little girl doll with unusual flirty sleep eyes in a mask face. Circa early to mid-1940s.

Many other cloth dolls were made by the Alexander firm during the 1930s and 1940s. With the company's ability to market a basic doll dressed in many different costumes, it may be impossible to ever identify all of the doll names.

Georgene Novelties, Inc.

The Georgene Novelties firm was founded in the early 1920s by Georgene Averill (see Georgene Novelties chapter). During the 1930s and 1940s, the Georgene Novelties company made many beautiful cloth dolls, specializing in "International Dolls" which were made in pairs. Most of these dolls were 12"-14" tall. The dolls are often confused with the same types of dolls produced by Molly'es but the early Georgene dolls featured real eyelashes while the Molly'es eyelashes were painted. Besides the smaller dolls, the Georgene firm also marketed some large 26" pairs of "International Dolls" circa 1930s. Two of these dolls, dressed as Dutch children, have been identified with original Georgene tags. The dolls had painted features, mask faces, yarn hair, and added eyelashes.

The Georgene firm was also famous for its Raggedy Ann and Andy dolls, which the firm produced for twenty-five years beginning in 1938 and ending in 1963 (see Character Dolls chapter).

Other very collectible Georgene dolls include the comic dolls: Little LuLu, Tubby, Alvin, Nancy and Sluggo (see Character Dolls chapter). These dolls are very hard to find in excellent condition.

More easily found Georgene cloth dolls include the Brownie and Girl Scout dolls circa late 1940s and the Topsy and Eva doll (see Georgene Novelties chapter). Another hard-to-find Georgene cloth doll is the Miss America doll. She wore a Statue of Liberty type headdress.

Besides international dolls, Georgene Novelties made a series of story book dolls featuring mask faces, painted features (no applied eyelashes), and floss hair. These dolls were 12-1/2" tall. Their clothes were fastened with round snappers.

Many unidentified cloth dolls may also have been made by the Georgene firm, as they were in business for so many years and the cloth dolls were some of their most popular and long-lasting products.

Ideal Toy Corporation

Although the Ideal Toy Corporation is most known for their composition dolls in the 1930s and early 1940s, the firm did market several very desirable cloth dolls.

The Story Book line of cloth dolls was advertised in the John Plain catalog in 1939. The advertisement pictured only "Mary Had a Little Lamb" and "Little Miss Muffet" but the series also included "Little Bo Peep, Queen of Hearts, Little Boy Blue, Tom, Tom the Piper's Son, Mistress Mary-Quite Contrary, and Daffy Down Dilly."

The Ideal catalog for 1939 pictured all of the dolls. They were described as being 16" tall, stuffed with kapok, with mask faces, washable hands, turning heads, and worsted hair. They were to retail for $2.00 each. The individual doll's character name was printed on the clothing, usually the apron.

At about this same time, the Ideal firm was also producing a series of Snow White dolls to tie in with the Walt Disney movie *Snow White and the Seven Dwarfs*. The company marketed Snow White dolls in both composition and cloth (see Character Dolls chapter). The Seven Dwarfs were also made in cloth and were 12" tall.

In the early 1940s, the Ideal firm produced another very nice cloth story book doll when they marketed Mary and Her Lamb. This doll had painted features with her mouth drawn in a circle—a popular "look" of the time. She was featured in the 1943 Sears Christmas catalog at a price of $5.98. A stuffed lamb was also included with the doll.

Other Ideal cloth dolls of the period included bunny dolls in 1941. These dolls came in sizes of 14" and 18" and had children's faces with bunny ears. "Betty Big Girl" was also advertised by Ideal in 1941 and "Bit of Heaven," a cloth doll with closed eyes and angel wings, was marketed in 1946. Two 40" cloth dolls, called Farmer Girl and Farmer Boy, were also advertised that same year.

Because the Ideal firm produced so many different styles of cloth dolls during the late 1930s into the late 1940s, it is likely that many other different dolls were marketed that have not been identified. Only tagged dolls or original advertising can identify additional dolls made by the Ideal company.

13-1/2" cloth international doll attributed to Georgene Novelties. She has a molded buckram face with painted features, yarn hair, and mitten hands. She is near mint and all original. She wears removable shoes with scalloped trim (seen on tagged Georgene cloth dolls). Her clothing is made of felt ($100+).

Two of Ideal's 16" cloth Story Book dolls were featured in the John Plain catalog in 1939. The same basic doll was used for the whole series of dolls; the clothing determined the identity of the dolls. These two examples are "Mary Has a Little Lamb" and "Little Miss Muffet." All of the dolls were identified through printing on their clothing, in these cases on their aprons. The dolls were made of a treated material and the company copy said their hands could be washed. They had mask faces, turning heads, yarn hair, and were stuffed with kapok. *From the collection of Marge Meisinger.*

Cloth Dolls

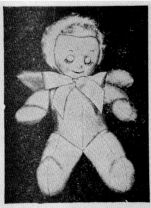

Left:
30" cloth doll which may have been made by Ideal, circa mid to later 1940s. The company marketed several versions of cloth dolls with O-shaped mouths. This doll has a molded painted buckram face, yellow yarn hair in braids, a separate neck piece so the head turns, and a painted O-shaped mouth. She wears her original red cotton dress and panties. Her red shoes are made as part of the body. She may have originally had an apron or pinafore over her dress ($75+ incomplete costume).

Right:
The Chicago Wholesale Company Fall and Winter catalog for 1946 and 1947 featured Ideal's 15" cloth "Little Bit of Heaven" doll. The copy described the doll as being made of fine quality rayon fleece. Her head covering and wings were made of pangora plush. Her eyes were painted shut. *From the collection of Marge Meisinger.*

Several Ideal cloth dolls were pictured in the Chicago Wholesale Co. catalog for Fall and Winter 1946-47. Included were 40" Farmer Girl and Farmer Boy dolls. They had pressed, hand painted mask faces and yarn hair. Both dolls had the O-shaped mouths and they appeared to be the same basic doll, dressed as both a boy and a girl. The clothing was removable but the shoes appeared to be made as part of the body. A 15" cloth little girl doll was also pictured. She had a hand painted mask face, yarn hair, removable clothing, and shoes made as part of the feet. *From the collection of Marge Meisinger.*

The Sears Christmas catalog circa 1943 featured a Mary and Her Lamb cloth doll thought to have been made by Ideal. The doll was 25" tall with a molded buckram face, painted features (mouth in an O-shape), and hair of soft rayon plush. Her arms and body match the skirt of her dress. A pinafore and bonnet complete the costume. She carried a lamb which came with the doll. Her shoes and socks were part of her body. She sold for $5.98 complete.

Mollye/International Doll Co.

The Philadelphia based Molly'es firm was very active in marketing cloth dolls in the 1930s and 1940s. Since the Mollye Goodman company was best known for its ability to dress dolls made by other firms, the dolls that were sold by Molly'es may have actually been produced by other companies.

One of their most popular lines was their series of international dolls. These cloth dolls are from 13" to 15" tall and are often confused with the dolls marketed by Georgene Novelties. The early Georgene dolls had real eyelashes while the Mollye International Dolls did not. The firm usually used mohair for the doll's wigs while Georgene dolls had yarn hair. All of the dolls had mask faces with painted features.

Cloth Dolls 137

For a few years (1935-38), Molly'es produced unauthorized Raggedy Ann and Andy dolls. (see Character Dolls chapter).

The firm also marketed many other cloth dolls, including an Uncle Sam in 1940. Other cloth dolls are hard to identify without the original tags but many more models must have been made.

11-1/2" cloth doll that may have been marketed by Molly'es. She has the same type long arms sticking out to the sides, eyelashes painted to the right of the eyes like the Molly'es Russian dolls, and machine stitching up the back like the Molly'es examples. She does have a yarn wig instead of mohair, but perhaps that is because she wears no covering on her head. She is one of the author's childhood dolls dating to 1943. She is all original except for a missing shoe ($60-75).

12-1/2" cloth International Russia doll marketed by Molly'es. She has a molded mask buckram face, painted features, mohair bangs, and mitten hands. She has a stuffed bust and is wearing a dress, slip, and cheap underpants, earrings, felt boots, and hat. She carries her original tag, which reads "MARYASSA OF RUSSIA/CREATED BY MOLLYE/INTERNATIONAL DOLL CO./Philadelphia, Penna." The back of the tag lists other dolls available, including England, Switzerland, France, Sweden, Holland, Mexico, and Czechoslovakia ($75-100).

8" original cloth doll, circa 1945, with molded mask face. She may be a Molly'es product as her arms stick straight out and her eyelashes are painted to the right as are the tagged Molly'es dolls. She also has the mohair bang with a cap covering the rest of the head ($75+). *Childhood doll of Bonnie McCullough. Photograph by Marilyn Pittman.*

Left:
12" cloth International Russia doll created by Molly'es. The molded mask buckram face of this doll is very different than the other Russia doll pictured. The face is much flatter, with the nose, eye sockets, and mouth moldings less distinct than on the other, more attractive Russian face. This may be because the Molly'es firm purchased the face masks from another company and used whatever was available. The same tags are on both dolls. The clothing on this doll seems to have been produced from odds and ends. Her underwear is made of flowered cotton flannel, the front and back of the blouse is made of a flowered cotton that does not match her skirt, and the sleeves of the blouse match the apron. The doll does wear real doll shoes. She still has her original box ($100).

Richard G. Krueger Inc. (R.G. Krueger)

The New York Richard G. Krueger firm was very active in making cloth dolls in the 1930s and 1940s. Many of the doll bodies were made of an oilcloth type material. Sometimes the feet were made of this material to indicate shoes. According to Polly Judd, writing in her book *Cloth Dolls*, many of the Krueger dolls had feet that turned up and came to a point.

The firm's most collectible dolls are the Snow White and Dwarfs set, circa 1940s. The dwarfs were 12" tall with bodies made of velveteen. Cardboard was used on the bottoms of the feet to make the dolls able to stand. They had mask faces and painted features.

Other collectible dolls include the Cuddle Kewpies produced in the late 1920s and early 1930s. They were made with jersey bodies, tiny wings, and a peak on the head. The firm also produced a cloth Scootles in 1935. This doll came in 10" and 18" sizes and had blonde curly hair and mitten hands. Both dolls were designed by Rose O'Neil.

Krueger also marketed a Pinocchio doll circa 1940 to tie in to the Disney movie of the same name. The doll had a mask face, cloth torso, black yarn hair, and wood jointed arms and legs.

It is known that the Krueger firm marketed many more cloth dolls during its time in business. Original advertising and tagged dolls are needed in order for collectors to identify more models.

9" Cuddly Kewpie designed by Rose O'Neill and thought to have been made by Richard Krueger in the 1930s. These dolls had painted flat faces, were jointed at the shoulders and hips, and had tiny wings on the back and a peak on top. The stuffed cloth dolls were made of jersey. They were sold for several years in different sizes ($75+ with wear).

12-1/2" cloth doll possibly made by Richard G. Krueger. His turned up feet, made as part of the body, are of an oilcloth type material. This is one of the characteristics of some of the Krueger dolls. The doll has a molded buckram face with painted features, yarn hair, mitten hands, and removable clothing trimmed in the same material as that used on the feet. He is probably missing his hat ($55-75).

Tiny Town Dolls

The Tiny Town Dolls' trademark was registered in 1949. Alma Le Blane doing business as Lenna Lee's Tiny Town Dolls in San Francisco, California held the trademark.

The cloth dolls were made in sizes from 3-1/2" to 7-1/4" tall. Most of the dolls have molded felt faces with painted features, mohair wigs, cloth wrapped armature bodies, arms, and legs, with felt mitten hands. Their shoes and socks are weighted molded metal. The larger 7-1/4" dolls have a different construction which includes regular arms and legs and the dolls wore removable shoes.

4" Tiny Town doll dressed as Red Riding Hood, circa 1949. She has a felt face, painted eyes, thread wrapped armature body, arms and legs, felt mitten hands, and molded metal feet and socks. She is all original with her basket and wrist tag. She wears a blue and white check dress with cape. The tag reads "Red Riding Hood" on one side and "Tiny Town Dolls" on the other ($75). *Doll from the collection of Marge Meisinger. Photograph by Carol Stover.*

The dolls were dressed in a variety of costumes including ballerinas, Red Riding Hood, Hansel & Gretel, and various school girl outfits. Like the Nancy Ann dolls, the only way the individual doll can be identified is from the original tag.

A set of dollhouse dolls, which included a father, mother, daughter, and son, was also marketed by the firm.

Since these dolls have just recently become popular with collectors, it is likely that many more dolls will be identified in the next few years.

4" Tiny Town ballerina Twinkie and two 5" ballerinas with felt faces, painted features, thread wrapped armature bodies, arms and legs, and felt mitten hands. They wear their original costumes featuring tulle skirts ($65-75 each). *Dolls from the collection of Marge Meisinger. Photograph by Carol Stover.*

4" Tiny Town dolls dressed as Swiss Boy and Girl, circa 1949. They have felt faces with painted eyes, thread wrapped armature bodies, arms and legs, and felt mitten hands. The shoes are molded on their metal feet. The tagged dolls are all original with costumes of cotton and felt ($135-150 pair). *Dolls from the collection of Marge Meisinger. Photograph by Carol Stover.*

Tiny Town dollhouse dolls, circa 1949. The dolls range in size from 3-1/2" to 5" tall. The bendable dolls have been made with armature construction and have hand painted features. The dolls have metal feet and socks that are made and painted in one piece. Their faces are sculptured and the hands look like mittens. Most of the original clothing is felt, although the little girl is wearing a cloth dress ($40+ each).

4" Tiny Town dolls dressed as Hansel & Gretel, circa 1949. They have felt faces with painted eyes, thread wrapped armature bodies, arms and legs, and felt mitten hands. Their original costumes are of cotton and felt with wood shoes, and they have their original wrist tags ($135-150 pair). *Dolls from the collection of Marge Meisinger. Photograph by Carol Stover.*

Miscellaneous

Many other cloth dolls remain to be classified as to maker. Some of these dolls were produced very cheaply while others were made and dressed with care and style. Only continued research can unlock the puzzle regarding identification of the many companies who marketed these cloth dolls of the 1930s and 1940s.

17" twin cloth dolls purchased in 1943. They have painted flat faces, hair made of a plush material, and bodies and limbs of red and white striped cotton. Both dolls are dressed in their original blue cotton outfits. Their red shoes are made as part of the body. They are childhood dolls of the author. It is thought that the dolls were priced at 75 cents each and were purchased at the Union Station in Kansas City, Missouri. The male doll had a second life with another generation and the affection shown reflects on the additional wear on his body. Maker unknown (sentimental value only).

N. Shure Co. advertised the 16" cloth Effanbee Pat-a-Pat doll in their 1941-42 catalog. The copy read that the doll had been designed for infants and toddlers. If the doll's stomach was pressed, she clapped her hands and cried. She had a mask face, wool hair, and a soft stuffed body. *From the collection of Marge Meisinger.*

Below:
18" unidentified cloth doll with molded mask face, sleep eyes, real eyelashes, yarn hair, and mitten hands. She may have been made by the same company who produced the doll advertised in the 1944 Sears catalog. Her shoes and socks have been replaced but her dress may be original ($125+ with sleep eyes).

The Sears 1944 Christmas catalog featured an unusual cloth doll with sleep eyes. She had a mask face, real eyelashes, yarn hair, mitten hands, and was 17" tall. She sold for $4.88. *From the collection of Marge Meisinger.*

13" unusual cloth doll with a composition face. The back of the head is cloth. She has painted features and molded hair. Her body and limbs are made of printed cloth and she wears a matching bonnet. Her skirt is removable. She is not marked ($40-50).

Cloth Dolls 141

The Fall and Winter Chicago Wholesale Company catalog of 1946-47 featured several cloth dolls, including Jitterbug Jane. The 32" doll was geared to teen-agers. *From the collection of Marge Meisinger.*

16" unidentified cloth doll, circa 1940s. She has a molded mask face, painted features, yarn hair, and is all original. Her clothing is especially nice for a cloth doll. Her "snow suit" is trimmed in felt and includes matching mittens. With her eyelashes painted to the right and the extra effort with her clothing, she could be a Molly'es product. Her shoes are regular doll shoes ($100+).

The 32" cloth Jitterbug doll as advertised in 1946-47. She is missing her scarf, otherwise she is all original. She has a mask face, painted features, mohair curls, hands with fingers indicated by stitching, shoes and socks made as part of the body. These large cloth dolls were popular with teen-age girls during the 1940s ($50+).

The Sears Christmas catalog for 1945 pictured two cloth dolls. Both dolls had mask faces, painted features, and yarn hair. The smaller doll on the left was 18" tall and sold for $2.98, while the larger 20-3/4" doll was priced at $3.99.

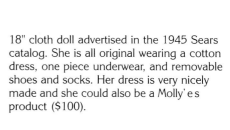

18" cloth doll advertised in the 1945 Sears catalog. She is all original wearing a cotton dress, one piece underwear, and removable shoes and socks. Her dress is very nicely made and she could also be a Molly'es product ($100).

142 Cloth Dolls

Left:
20" unidentified cloth doll with pigtails, circa 1940s. She has a buckram face, painted features, yarn hair, and mitten hands with more of a thumb indicated than usual. Her hair is only around the edges—her hat covers the rest of the head where there is no hair. She has felt shoes and removable clothing. She was purchased at an auction of old store stock from a hardware store that had gone out of business. She and other dolls had been stored in the attic for many years. The doll was still in her box but it gave no information as to maker (MIB $150+).

Right:
12" cloth Elceea topsy-turvy doll, circa 1940s. The two-headed doll has mask buckram faces, printed features, and hands sewed to indicate fingers like those of the Jitterbug doll. The doll has a tag that reads "ELCEEA." She is all original with a Dutch costume on one end and a sleeping baby outfit on the other ($75+).

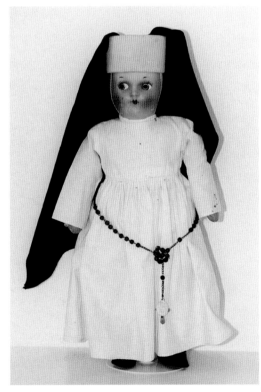

14" cloth Nun doll, circa late 1940s. Nun dolls seemed to be quite popular in the 1940s and early 1950s. Dolls of composition and hard plastic were also marketed dressed as nuns. This doll has a molded buckram face, oilcloth-like arms, mitten hands, and real feet with shoes. The oilcloth type arms might indicate a Richard G. Krueger connection. She is all original with her Rosary ($125+).

12" unidentified cloth international type doll, circa 1940s. She has a molded buckram face, painted features, mitten hands, and is all original. She is not tagged, so the country she represents is not known ($40-50).

Military and War Related Dolls

When the Japanese bombed Pearl Harbor on December 7, 1941, war was soon declared against Japan, Germany, and Italy. As the conflict escalated, the people of the United States came together in a united war effort never matched before or since. World War II dominated the lives of everyone for nearly four years. Of course, most of the younger men served in the armed forces, and for the first time in war, many women also joined the new female W.A.V.E, W.A.A.C, or W.A.A.F branches of service. Other women signed on as Red Cross volunteers or as nurses. The war plants were staffed with men not qualified for military service due to age, health concerns, or disabilities. To add more workers, women were encouraged—for the first time—to apply for work in defense plants. These women took over jobs that had been left vacant by men drafted into military service.

Dolls and toys of the era reflected the times as tanks, ships, and planes played a dominant role in the toy industry. Dolls and paper dolls, dressed in military and nursing attire, were popular with little girls. Both composition and cloth dolls were produced dressed in military styles. Because of their historical significance, these dolls are especially collectible today.

The following companies marketed composition dolls of this type:

Alexander Doll Co.: Following their usual practice, the Alexander firm used their basic Wendy Ann dolls and changed identification by adding unique costumes. The company marketed dolls dressed as a W.A.V.E, a W.A.A.C, a soldier, and a marine. The same basic doll was used for both the male and female models. Most dolls were 14" or 15" in size.

14" composition W.A.V.E. by Madame Alexander, circa 1942. She has sleep eyes, closed mouth, mohair wig, and is fully jointed. Her uniform includes a cotton twill suit with brass buttons, and leatherette purse and shoes. Her uniform jacket was tagged "W.A.V.E." She is all original. The doll was a basic Wendy Ann (not enough examples to determine a price). *Doll and photograph from the collection of Nancy Roeder.*

Left:
15" composition W.A.A.C. by Madame Alexander, circa 1942. She has sleep eyes, closed mouth, mohair wig and is fully jointed. Her uniform includes a cotton twill suit with brass buttons, and leatherette purse and shoes. Her uniform jacket was tagged "W.A.A.C." She was advertised in the Alexander catalog for 1942-43. She is all original. The doll was a basic Wendy Ann (not enough examples to determine a price). *Doll from the collection of Marge Meisinger. Photograph by Carol Stover.*

Right:
15" composition marine by Madame Alexander, circa 1943. She has sleep eyes, closed mouth, mohair wig, and is fully jointed. She wears a navy blue twill jacket accented with brass buttons, a black belt, navy pants with a red stripe, and suspenders. Her matching hat and black leatherette shoes complete her uniform. The doll is marked "Mme. Alexander" on the head and with a "Marine" tag on the jacket. It is hard to imagine this very feminine looking Wendy Ann representing a U.S. Marine (sold at auction for $945). *Courtesy of Frashers' Doll Auctions, Inc.*

Military and War Related Dolls

Effanbee Doll Co.: The Effanbee Doll Co. applied the same marketing strategy as Alexander. The firm used the Skippy doll already in production and dressed him in various armed service uniforms in order to compete in the military doll category. He was marketed costumed as a sailor, a soldier, and an aviator. Effanbee also issued their Little Lady dolls dressed in nurse outfits or in U.S.O. khaki overalls while Suzanne was marketed as a nurse or a W.A.A.C.

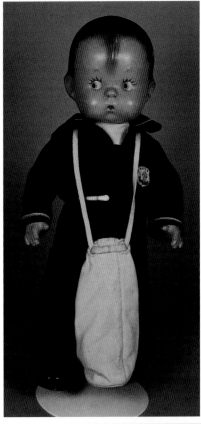

14" composition Skippy by Effanbee, circa 1943. He has painted eyes, molded hair, closed mouth, cloth torso and upper legs, molded composition lower legs to above the knee, black boots and socks, and composition arms. He is marked on the head "Effanbee/Skippy/©." He wears his original cotton twill sailor suit with white braid and brass buttons. The pin is not original to this doll. He is missing his hat. This doll was pictured in the Montgomery Ward Christmas catalog in 1943 and sold for $2.29 ($450+). *Doll from the collection of Marge Meisinger. Photograph by Carol Stover.*

Freundlich, Ralph A.: Some of the nicest military dolls of World War II were produced by the Freundlich firm. Their service related composition dolls featured molded hats along with removable uniforms. The most collectible of these dolls is the 18" General Douglas MacArthur doll (see Personality Dolls chapter). The company's soldier, sailor, W.A.A.C, and W.A.V.E dolls were 15" tall.

15" composition Freundlich sailor and W.A.V.E, circa 1942. Both dolls have painted features and molded hair and hats. They are all composition and jointed at the shoulders and hips only. The dolls are original except for the sailor's shoes. The *We Fly for the Navy* coloring book was published by Merrill Publishing Co. in 1943 (coloring book, $20-30, dolls $200 each).

Ideal: The Ideal company issued their 13" Flexy dolls in both sailor and soldier uniforms. These dolls had composition heads and hands and flexible wire arms and legs. Their torsos and feet were wood.

14" Effanbee Skippy, circa 1943. He has painted eyes, molded hair, closed mouth, cloth torso and upper legs, molded composition shoes and socks, and composition arms. Marked on his head is "Effanbee/Skippy/©." He is all original, wearing his cotton twill army uniform consisting of a jacket, shirt, tie, pants, and cap. A similar doll wearing an officer's cap was shown in the 1943 Montgomery Ward Christmas catalog. The *Fighting Yanks* coloring book was published by The Saalfield Publishing Co. during the World War II years (coloring book, $20-30, doll $450+).

13" Ideal composition sailor, circa World War II years. He has painted eyes, closed mouth, molded hair and is fully jointed. Marked "Ideal" on the back of his neck. A similar head was used on a flexy body but this doll is all composition. He is all original, including his sailor hat ($150).

Reliable Toy Co.: The Reliable Toy Co. of Toronto, Canada also offered a line of war related dolls. They included an 18" R.A.F. flyer and an 18" nurse doll. Both dolls have composition heads and arms, painted features, molded hair, and excelsior stuffed bodies. They are marked "Reliable/Made in Canada."

Vogue Doll, Inc.: Vogue Dolls, Inc. marketed more World War II related dolls than the larger, more well-known doll companies of the era. They, too, used dolls already in production and changed their identities through the use of costumes. Their 8" all composition Toddles doll was sold dressed as Uncle Sam, a Navy Officer, Air Force Officer, Sailor, Aviator, Soldier, War Nurse, Air Raid Warden, and a Miss America dressed in red, white, and blue. The company also issued 13" all composition W.A.V.E and W.A.A.C dolls in 1943 and 1944 called WAVE-ette AND WAAC-ette. The dolls had sleep eyes and mohair wigs. Their uniforms came complete with hats and shoulder bags. Many of these dolls were marked only with a round silver Vogue sticker.

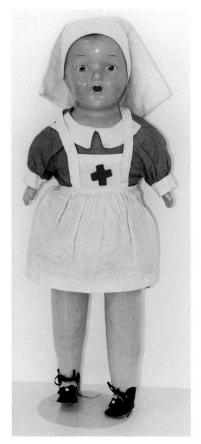

18" Reliable nurse, circa World War II years. She has painted eyes, closed mouth, molded hair, a composition shoulder head and short arms, and a cloth body and legs. She is marked "A/RELIABLE/DOLL/MADE IN CANADA." She is wearing her original nurse uniform but is missing her original shoes and perhaps socks ($100).

8" composition Vogue Toddles Uncle Sam from Military Group, circa 1943. He has painted eyes, closed mouth, mohair wig, and is fully jointed. He is marked "DOLL CO." on the back. He is all original wearing a striped cotton pant suit with blue cotton jacket and tails, a white felt hat trimmed with cotton stripes (the blue starred hat band has faded to gray), and red snapped shoes. A paper gold Vogue tag is on the pants. The doll has a bent right elbow ($300+). *Doll and photograph from the collection of Carol Stover.*

8" Vogue composition Toddles "Draf-tee," circa 1943. He has painted eyes, closed mouth, mohair wig, and is fully jointed. The doll is marked "DOLL CO." on the back. He is dressed in a brown flannel uniform and matching cap with a leatherette brim, leather belt, and shoulder strap. The brown leather side snap shoes have "DRAF-TEE" stamped on the bottom of the left shoe. A toy metal gun hangs on his side. The doll has a bent right arm and is in a Vogue labeled box ($400+ in box). *Doll and photograph from the collection of Carol Stover.*

146 Military and War Related Dolls

Miscellaneous: Many other lesser known companies issued dolls related to World War II. Most of these dolls were unmarked. A 13" all composition unmarked sailor with painted features and molded hair may have been issued by Ideal. It is likely that a soldier was also made in this line of dolls. A 9" all composition W.A.A.C doll also came with no identification mark. She has painted features and an applied mohair wig. Catalogs of the era picture other military and nurse dolls. The Carson Pirie Scott & Co. catalog from 1941 advertises an all composition 15" soldier with molded hair and painted eyes and an all composition 16" nurse with a mohair wig. The Spiegel 1941 catalog offered a most unusual composition doll for $2.29. The 16" doll came with three outfits so it could be dressed as a nurse, sailor, or soldier. Although changing the same doll from a male to a female seems strange, it was no different than the popular Alexander Doll Co.'s practice of using its basic Wendy Ann doll as both a soldier and a W.A.A.C, depending on its costume. Other companies simply dressed their dolls in red, white, and blue costumes in order to tie in to the patriotic feelings felt by nearly everyone in the United States during the war years.

9" unidentified composition W.A.A.C from World War II years. She has painted eyes, closed mouth, mohair wig, and is jointed at the shoulders and hips only. She is wearing her original uniform, complete with cap, and has molded shoes. She is pictured with a military paper doll book of the period (paper doll book, $50+, doll $40+).

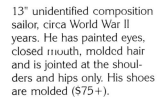

13" unidentified composition sailor, circa World War II years. He has painted eyes, closed mouth, molded hair and is jointed at the shoulders and hips only. His shoes are molded ($75+).

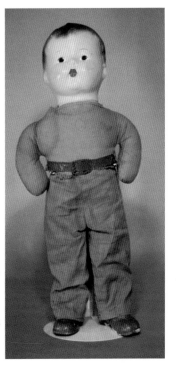

14" unidentified soldier, circa World War II years. He has painted eyes, closed mouth, molded hair, composition head, cloth body, arms and legs, and molded shoes. He is probably missing a hat and perhaps trim on his upper body, which serves as the top part of his clothing ($50).

The Carson Pirie Scott and Co. catalog from 1941 featured two war related dolls. The composition nurse was 16" tall and came with a white uniform and blue cape. She had painted eyes and a mohair wig. The composition soldier looks very much like the unidentified 13" sailor. He was 15" tall and wore a Khaki uniform and cap. He had painted eyes and molded hair. *From the collection of Marge Meisinger.*

Military and War Related Dolls 147

Left:
The 1941 Spiegel catalog advertised an unusual Three-In-One Doll. The 16" composition doll had sleep eyes and molded hair. The doll came with three outfits so it could assume the identity of a soldier, nurse, or a sailor. The outfit sold for $2.29. *From the collection of Marge Meisinger.*

Right:
7" composition doll, circa 1943. She may have been made by Beehler Arts Co. The doll has painted eyes, closed mouth, and a mohair wig. She is jointed at the shoulders and hips. Her original box with its V for Victory design and her patriotic costume tie in to the World War II years. Marked "MBC" on her back ($35+).

Organization Dolls: Many individual organizations dressed dolls to represent nurses or members of the armed forces to promote World War II causes and charities. In small towns, many of these dolls were displayed in store windows in order to solicit funds for various causes like "American War Mothers" or the "American Red Cross."

Left:
23" unidentified composition nurse, circa World War II years. She has flirty sleep eyes, open mouth with teeth, mohair wig, a composition shoulder head with shoulder plate, full composition arms and legs, and a cloth body. Her costume is not factory made. She has been dressed in a taffeta nurse uniform, which includes a dress, cape, apron, and head piece. Her shoes and socks are from a later period. Her "V for Victory" pin dates the outfit to World War II. It is possible that the doll was used to raise money for the Red Cross either in a store window or as an auction item ($150-200).

Right:
13" rare 11-piece World War II Red Cross doll set. Included are nine composition sewing mannequin dolls dressed in various Red Cross outfits, a cloth soldier in a metal wire wheelchair, and WWII Red Cross flag. Only the arms are jointed. This set may have been used during World War II to advertise the Red Cross and help in fund raising (sold at auction for $715). *Courtesy of Cobb's Doll Auction.*

Cloth dolls were also produced during World War II to tie in with the war effort. Some of these dolls include the following:

Dollywood Defense Dolls: Dollywood Defense Dolls were made by Dollywood Studios, Inc. in Hollywood, California during World War II. The all cloth dolls came in two sizes, 12" and 16". There were at least five designs of dolls, including a soldier, sailor, army officer, naval officer, and nurse. An advertisement featured in *Playthings Magazine* in 1942 said they were designed by one of America's leading artists.

Prager & Rueben: The Raggy-Doodle U.S. Parachute Trooper was marketed in 1942. The doll had a pack on his back where the parachute was packed. This doll was copyrighted in 1942 by M. Hoyle and was produced by Prager & Rueben of New York City. The English Norah Wellings firm made a similar doll during the war years.

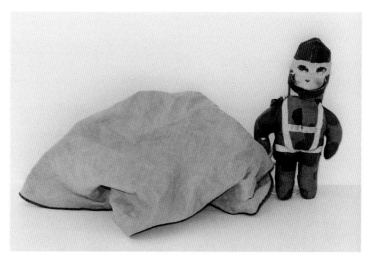

6-1/2" cloth Raggy Doodle US Parachute Trooper produced by Prager and Rueben in 1942. The doll's features were painted on molded reinforced cloth. The goggles are also cloth. His uniform was part of the body and he came with a pack on his back to hold his parachute. He is missing a fur piece around his neck ($35+).

12" and 15" cloth Dollywood Defense Dolls, circa 1942. The dolls have painted features, mitten shaped hands, and are made of cloth. They are all original, with the sailor 12" tall and the soldier 15" high. Both dolls have one arm tied to the head in a salute. The clothes are part of the bodies except for collars and trim ($50-75 each).

Other simpler cloth military dolls were made with their clothes as part of the body and a hat or accessory added for trim. Many of these dolls had buckram faces.

Most of the war related dolls were sold during 1942 and 1943. By 1944, Americans wanted relief from the emotional stress of the war and World War II related toys were no longer in favor. People were focusing on "after the war" and getting "back to normal." Toys, music, and entertainment reflected this attitude as most dolls and other products no longer used the war theme to sell merchandise. Since these "war" dolls were produced for only a few years, they are not plentiful, and their inclusion can add an interesting historical perspective to any doll collection.

Left:
This 14" composition unmarked doll models a sailor uniform made by the owner's mother when she was a child during World War II. It was made to fit an R&B boy doll that was already a childhood toy. The outfit is still in the collection of the original owner. *Doll, outfit, and photograph from the collection of Nancy Roeder.*

Right:
16" cloth soldier, circa World War II. The doll has painted features on a molded buckram face and yarn hair. His hands and feet are made of a plush-like material and the body, arms and legs are corduroy. The added trim is felt and gold. This doll was also produced as an aviator ($50+).

Personality Dolls

Shirley Temple

In 1934, during the midst of the worst economic depression this world has ever known, America fell in love with a tiny golden haired moppet: Shirley Temple. Now, nearly seventy years later, some people are still actively pursuing their love affair with Shirley by collecting Shirley Temple memorabilia.

Shirley was born on April 23, 1928 in Santa Monica, California. Because she loved to dance to the radio music, her mother enrolled her in a dancing school. While she was there, Shirley was seen by a movie scout and was signed to make short films for the Educational Studios.

Shirley's charm in these movies caught the attention of Leo Houch, assistant director for Fox Film Corp. and he gave her a starring role in *Stand Up and Cheer* in 1934. Shirley's golden curls and her pleasing personality made her an instant hit with the public, so she was signed to a seven year contract with Fox.

In 1934, Shirley made twelve pictures, many of them still rated as some of her best. These include: *Little Miss Marker* (Paramount), *Baby Take a Bow, The Little Colonel*, and *Bright Eyes,* all for Fox.

There has never been a child movie star as popular as Shirley Temple. Her good looks and talent made her a perfect role model for little girls all over the world. Mothers began shaping their own little girls' hair into ringlets in an attempt to copy Shirley's curls. Dresses, toys, dolls, and many other products were manufactured to take advantage of the Shirley name (see *Hollywood Collectibles* by Dian Zillner for more information and pictures of other products).

The most desirable of the Shirley endorsed toys are the Shirley Temple dolls, which have been made in several different designs through the years. The first dolls were produced of composition by the Ideal Novelty and Toy Co. These dolls were marketed from 1934 to 1940 and came in many sizes, including 11", 13", 15", 18", 20", 22", 25", and 27". The dolls had sleep eyes, open mouths with teeth, and beautiful mohair wigs featuring the famous Temple curls. Because the dolls were rather expensive, other doll companies also made imitation Shirley Temple dolls, but the dolls made by Ideal and the Canadian Reliable firm were the only ones authorized by Shirley Temple.

Although most Shirley Temple dolls were little girl models, the Ideal Co. also made a Baby Shirley Temple. This doll had a composition head, arms and legs, and a cloth body. It came with either a blonde mohair wig or molded hair, sleep eyes, and an open mouth. It was made in a variety of sizes, including 16", 18", 20", and 25". The most desirable of the Shirley Temple dolls are the mint dolls, with original clothing, pins, and boxes. These dolls are, of course, hard to find and therefore bring a top price.

By 1935, Shirley Temple was in the number one position among the top ten box office stars of the country. She retained that spot through 1938.

Popular Shirley films from the 1930s include *Our Little Girl, The Littlest Rebel, Curly Top, Captain January, Poor Little Rich Girl, Stowaway, Dimples, Wee Willie Winkie, Heidi, Rebecca of Sunnybrook Farm, Just Around the Corner, Little Miss Broadway, The Little Princess*, and *Susannah of the Mounties*.

By 1940, Shirley had lost some of her cuteness and her last two movies for Fox were not very successful, so her contract with Fox was ended. She graduated from high school in 1945 and married Sgt. John Agar on September 19, 1945. A daughter, Linda Susan, was born to the couple in 1948.

Shirley continued making films for various companies through the 1940s but never again received the stardom she had known as a youngster. Perhaps the most successful movie she made during this period was *Fort Apache* in 1948, which starred John Wayne and Shirley's husband John Agar.

Shirley retired from pictures in 1950. After divorcing Agar, she married Charles Black and changed her life to become a full time wife and mother. Two more children joined the family in the 1950s, Charles in 1952 and Lori in 1954.

In 1957, Shirley Temple came out of retirement and agreed to do the NBC *Shirley Temple Storybook* series on television. With the publicity generated by the show, the child Shirley Temple had been was again in the spotlight. Many products were produced to take advantage of this publicity. The Ideal vinyl dolls are especially attractive (see *Dolls and Accessories of the 1950s* for more information and pictures).

During the 1960s, child star Shirley Temple surprised everyone by becoming active in politics. From her home in California she ran an unsuccessful Republican campaign for Congress in 1967. After she helped Richard Nixon campaign for president, he appointed her as a U.S. delegate to the United Nations. Shirley was later appointed to several Ambassador jobs and served the nation well.

Unlike so many child stars, Shirley Temple has had a very successful life, even after the cameras stopped rolling. Perhaps that adds to the interest collectors still have in the memorabilia associated with the young star. The composition mint dolls continue to top the list of desirable Shirley Temple collectibles.

15" composition Shirley Temple by Ideal Toy Co., circa 1936. She has sleep eyes, open mouth with teeth, original mohair wig and is fully jointed. Both head and body are marked "Shirley Temple." She wears her original velveteen coat and hat ensemble with red buttons ($750). *Doll from the collection of Marge Meisinger. Photograph by Carol Stover.*

Personality Dolls

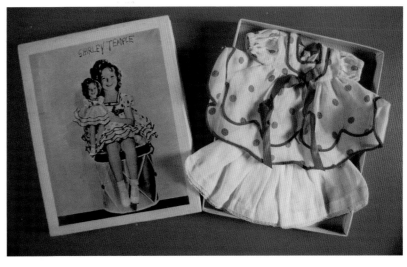

Boxed Shirley Temple coin dot dress from the 1934 movie *Stand Up and Cheer* with a slip-panty undergarment. Both items are tagged size 14 and were made by the Ideal Novelty and Toy Company in the 1930s ($200+). *From the collection of Mary Stuecher. Photograph by Werner Stuecher.*

Left:
The Sears winter catalog for 1935-36 carried ads for both the Shirley Temple little girl dolls and the Shirley baby dolls. The girl dolls came in sizes of 13", 16", 18", and 20". The baby dolls were in 15-1/2", 18", and 20" sizes. The dolls were priced from $2.89 to $5.79 each. Dresses for the dolls could also be purchased for 94 cents to $1.59.

21" Ideal composition baby Shirley Temple doll, circa mid 1930s. She has flirty sleep eyes, open mouth with teeth, and a mohair wig. She has a composition swivel head on a composition shoulder plate, composition arms and legs, and a cloth body with a crier. The back of her head is marked "Shirley Temple." She is all original wearing her tagged dress, bonnet, underwear, shoes and socks ($1,000+). *Doll from the collection of Lois Jakubowski. Photograph by Robert Jakubowski.*

25" Ideal composition baby Shirley Temple doll, circa mid 1930s. She is all original and in excellent condition. These dolls were also made with molded hair ($1,400+). *Doll from the collection of Lois Jakubowski. Photograph by Robert Jakubowski.*

Personality Dolls 151

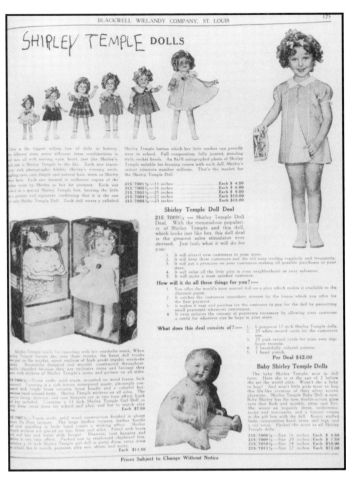

The Blackwell Wielandy Co. featured Ideal Shirley Temple dolls in their catalog for 1935-36. Pictured are the little girl dolls in sizes of 13", 16", 20", 22", and 25". The baby dolls with molded hair came in sizes of 16", 20", 24", and 27". Shirley Temple trunk outfits that included a 13" Shirley doll plus clothing were also shown.

15" composition Ideal Shirley temple from the movie *The Little Colonel* in 1934. She is all original with her box, pin, and NRA (National Recovery Act) tagged dress ($850+). *Doll from the collection of Marge Meisinger. Photograph by Carol Stover.*

These two Ideal composition Shirley Temple dolls are wearing costumes from the 1934 Fox film, *Bright Eyes*. They are 13" and 25" tall and are all original (13" $650-750, 25" $1,000+). *Dolls from the collection of Lois Jakubowski. Photograph by Robert Jakubowski.*

20" Ideal composition Shirley Temple wearing musical note dress from *Our Little Girl* (1935). She has sleep eyes, open mouth with teeth, a mohair wig, and is fully jointed. The dress, head, and body are marked "Shirley Temple." She is all original including her pin ($900+). *Doll from the collection of Lois Jakubowski. Photograph by Robert Jakubowski.*

13" and 15" Shirley Temples wearing "Scotty" dresses from the 1935 film, *Our Little Girl*. The dolls are all original (13" $650+, 15" $750+). *Dolls from the collection of Lois Jakubowski. Photograph by Robert Jakubowski.*

20" Ideal Shirley wearing a dress from the 1935 Fox film, *Curly Top*. She is all original complete with her original box and pin. Her box still shows the original price of $10.50, which was a good deal of money in the Depression years (boxed $900+).

1936 Shirley Temple Texas Rangers marketed in honor of the 1936 Texas Centennial. The 11" and 17" dolls are all original complete with their pins and guns (11" $850, 17" $800). *Dolls from the collection of Lois Jakubowski. Photograph by Robert Jakubowski.*

22" Ideal Shirley wearing a dress from the 1935 film, *Curly Top*. She is all original ($800+). *Doll from the collection of Lois Jakubowski. Photograph by Robert Jakubowski.*

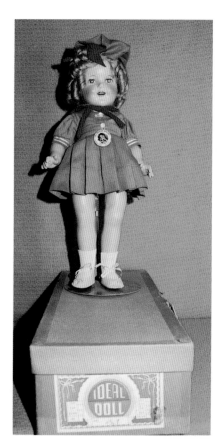

18" Shirley doll wearing a costume from the 1936 *Poor Little Rich Girl* film. She is original with her pin and box ($1,000+). *Doll from the collection of Lois Jakubowski. Photograph by Robert Jakubowski.*

Personality Dolls 153

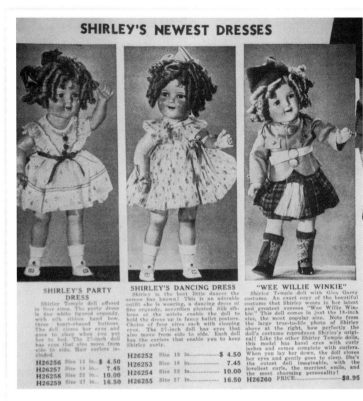

The John Plain catalog for 1938 was still featuring Ideal composition Shirley Temple dolls. Included was a doll costumed in an outfit from her *Wee Willie Winkie* film of 1937. It was 18" tall. *From the collection of Marge Meisinger.*

18" Wee Willie Winkie doll as pictured in the John Plain catalog. She is all original with her pin ($800+). *Doll from the collection of Lois Jakubowski. Photograph by Robert Jakubowski.*

18" Ideal composition Shirley wearing a tagged costume from the *Heidi* film of 1937. Instead of curls, she has a mohair wig styled in braids. She is probably missing her hat and may have originally come with wooden shoes. This outfit is hard to find ($800+). *Doll from the collection of Marge Meisinger. Photograph by Carol Stover.*

18" Ideal composition Shirley Temple wearing a costume from the 1938 Fox film, *Rebecca of Sunnybrook Farm*. She is all original, including her pin ($800). *Doll from the collection of Lois Jakubowski. Photograph by Robert Jakubowski.*

154   Personality Dolls

The Sears Christmas catalog for 1936 pictured several Shirley Temple items in a full page advertisement. Included was a new 11-1/2" size Shirley doll for $2.19. Other size dolls included 13", 16", 18", and 20". The largest doll sold for $5.79. Two styles of Shirley doll buggies were also pictured. The wood model was priced at $5.39 and the woven fiber buggy cost $7.98. Both models featured Shirley's picture on the sides. A Shirley trunk set was also offered. It cost $4.79 and included a 13" doll, party dress, play dress, bonnet, pajamas, hangers, and, of course, the trunk. *From the collection of Linda Boltrek.*

13" Shirley doll pictured with a Shirley Temple trunk and her homemade clothing. The trunk hangers are not original. The doll wears her original dress from *Curly Top* (doll $500, trunk $175-200).

Shirley Temple fiber doll buggy, 29" tall, made by the F.A. Whitney Carriage Co. circa 1936. There is a metal medallion with Shirley's picture on it attached to the buggy sides and the hub caps have "Shirley Temple" in script written on them. Inside the buggy is a very rare cloth-bodied Shirley Temple composition doll made by Ideal in the 1930s. The head is marked "Shirley Temple © Ideal N&T Co." The hairstyle is the later one with a smooth top and curls pulled to the side that Shirley first wore in *Rebecca of Sunnybrook Farm* (buggy $600). *Doll and buggy from the collection of Mary Stuecher. Photograph by Werner Stuecher.*

Personality Dolls

13" Ideal composition Shirley Temple doll mounted on a musical turntable. The doll turns as the music is played (not enough examples to determine a price). *Doll from the collection of Lois Jakubowski. Photograph by Robert Jakubowski.*

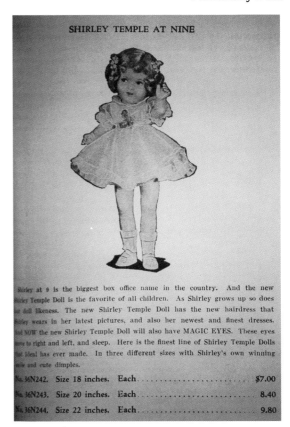

Ideal Shirley Temple dolls, with a new hairstyle, were advertised in the N. Shure Co. catalog of 1938. This design was flat on top and pulled back on the sides. The new doll also featured MAGIC EYES, which most companies called flirty eyes. The doll came in sizes of 18", 20", and 22". *From the collection of Marge Meisinger.*

Left:
18" Ideal composition brown Hawaiian doll using the Shirley Temple doll molds and marked "Shirley Temple" on her back. She has painted eyes, black yarn hair, and is fully jointed. The doll was not meant to represent Shirley Temple and was perhaps a tie-in to the popular *Hurricane* movie from 1937. She is all original ($900). *Doll from the collection of Lois Jakubowski. Photograph by Robert Jakubowski.*

Right:
22" composition Shirley Temple doll made by the Reliable Toy Co. of Canada, circa 1935-36. In addition to Ideal, this firm was also allowed to make authorized Shirley Temple dolls. They probably used the same molds as those used by Ideal. The dolls had sleep eyes, mohair wigs, open mouths with teeth, and were fully jointed. This doll is all original. The dress tag reads "A GENUINE/SHIRLEY TEMPLE/DOLL DRESS/RELIABLE TOY CO. LTD/MADE IN CANADA." ($1,200+). *Doll from the collection of Marge Meisinger. Photograph by Carol Stover.*

Dionne Quints

The Dionne Quintuplets were born in Callander, Ontario, Canada on May 28, 1934, in a small farmhouse, to rural parents with five other children. Two midwives and the local doctor, Dr. Allan Roy Dafoe, delivered the babies. The little girls were so small they were not expected to survive. When the astonishing news of their birth reached the outside world, the assistance that came from many companies and individuals helped save their lives. The little girls (Yvonne, Cecile, Emilie, Annette, and Marie) began to thrive after the Red Cross took over their care, which included payment for nurses and supplies. In order for the family to continue to receive help with the children, Elzire and Oliva Dionne were forced to sign over guardianship of the infants to the government. With donated funds, a new facility was built to house the babies along with their nurses, and the Quints were moved into the new nursery in September 1934. A playground was added to the structure so that visitors could watch the children through a screen, and a new tourist industry for Canada was established. From 1934 to 1943 around three million people visited the Quints in Callander.

In 1935, the Canadian government passed the Dionne Act of Parliament, which made the girls wards of His Majesty the King until the age of eighteen. A Board of Guardians was appointed, whose responsibility it was to look after the Quints' interests. Dr. Dafoe was made one of those guardians.

Product endorsements from the Quints began in 1935. The Madame Alexander Doll Co. first started producing the Dionne Quintuplet dolls in that year. The Alexander Company marketed over thirty different sets of the dolls with both bent legs and straight legs. They ranged in size from 7" to 23" tall. Although most of the dolls were all composition, several were produced with cloth bodies and composition limbs. These included both baby and little girl models. Various designs of the dolls continued to be produced until 1939. The later dolls were little girl dolls in keeping with the Dionne Quints' real ages. Besides Dionne dolls, the company also included a nurse doll with some of the sets of baby dolls as well as a unique 14-1/2" Dr. Defoe doll in 1936. This doll represented the doctor who had delivered and cared for the Quints. He was made of composition and featured painted eyes and a grey wig. He was dressed in a doctor's smock.

Accessories were also marketed by Alexander to accompany the dolls, including beds, play pens, high chairs, kiddie cars, and a chair. A Ferris wheel was also produced, filled with Dionne dolls.

The only other doll firm allowed to make authorized editions of the Quints was the Superior Co., based in Canada. Companies who made similar but unauthorized dolls included Arranbee, Freundlich, Effanbee, and a Japanese firm.

Many other Quint tie-in products were marketed in the 1930s and early 1940s. Included were paper dolls, books, hankies, toy dishes, post cards, souvenir items, advertisement calendars, spoons, fans, and many more items (see *Hollywood Collectibles* by Dian Zillner for more information and product pictures).

During these years, the Quints appeared in three movies for Twentieth Century-Fox: *The Country Doctor* (1936), *Reunion* (1936), and *Five of a Kind* (1938). More memorabilia is available for collectors from these motion pictures.

Since the parents had no control over the business matters of the girls, they were powerless to stop any enterprises they did not like. As the years passed, conflict continued to mount among the doctor, the government, and Mr. and Mrs. Dionne. After the Dionne's lawsuits brought against Dr. Dafoe (partly because of his financial benefits from the Quints) were settled out of court, the doctor resigned from the Dionne Quints' guardianship board in 1939. The problems between the two factions were still not resolved and in 1942 Dr. Dafoe resigned his position as the Quints' physician. He died in 1943.

The Dionne parents had been trying to regain control of the Quints since the girls were infants and finally were able to reunite their family in 1943. A big house was built to accommodate all of the Dionne children. The large family consisted of the Quints and seven other children (one more child was born in 1946). The Quints never seemed to "fit in" and they did not make an easy adjustment. Emilie died in 1954 of an epileptic seizure and Marie died in 1970, apparently from a clot to the brain. Yvonne passed away in 2001, at the age of sixty-seven. Although Marie, Cecile, and Annette married and had children, the marriages all ended in divorce. The two remaining sisters, Cecile and Annette, live just outside Montreal. The girls and their parents never reconciled after a dispute over a book the Quints helped write about their early lives.

Left:
The 1935 Montgomery Ward Christmas catalog featured the Dionne Quint dolls on the cover. Several different sizes of dolls were listed for sale, including the 7-1/2" composition baby dolls for 79 cents each, 10-1/2" baby dolls for $1.69 each, 17-1/2" baby dolls for $2.69 each, and 23" baby dolls for $5.79 each. A set of the small dolls sold in a doll bed for $4.39.

Right:
Dionne Quint doll as featured on the cover of the 1935 catalog. The doll is 17-1/2" tall with a composition head, arms and legs. She has sleep eyes, closed mouth, and molded hair, a cloth body and composition shoulder plate. The doll was made by the Madame Alexander Doll Company. She is wearing her original clothing but is missing her bib. The tag on her dress reads "Dionne Quintuplets/ Yvonne/Exclusive Licensee/Alexander Doll Co., N.Y." On the back of the doll's neck is "Dionne/Alexander." ($325+).

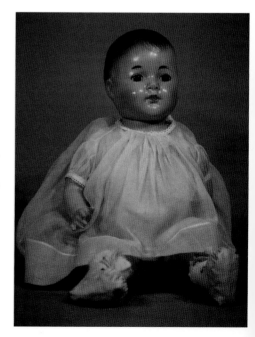

Personality Dolls 157

The Marshall Field & Co. Chicago catalog featured the 11" tall Alexander Quint baby dolls in their 1935 catalog. The dolls were composition with sleep eyes, closed mouths, and molded hair. The babies had their names embroidered on their bibs. They were Marie, Yvonne, Emilie, Cecile, and Annette. *From the collection of Marge Meisinger.*

The John Plain catalog for 1937 also offered a set of the toddler Quints in a 7-3/4" size. The composition dolls could be purchased as a group or individually. Their names were engraved on gold finished lockets they wore around their necks. The dolls had molded hair. *From the collection of Marge Meisinger.*

8" Alexander composition toddler Dionne Quints as pictured in the John Plain catalog for 1937, except with mohair wigs instead of painted hair. The dolls have painted eyes, closed mouths, and are fully jointed. They are dressed in their original Alexander clothing but are missing their lockets, shoes, and socks. They are marked "Dionne/Alexander" on their heads and "Alexander" on their backs. The clothing tags read "GENUINE/Dionne Quintuplet Dolls/ALL RIGHTS RESERVED/Madame Alexander-N.Y." (set $1,000-1,200).

7-1/2" Alexander composition Cecile Dionne Quint. She has painted features, closed mouth, molded hair, curved baby legs, and is fully jointed. She is all original including her "Cecile" bib. Her tag reads "This is/Cecile/of the/DIONNE QUINTUPLETS" and "Created/by/Madame Alexander/New York." ($225+).

158 Personality Dolls

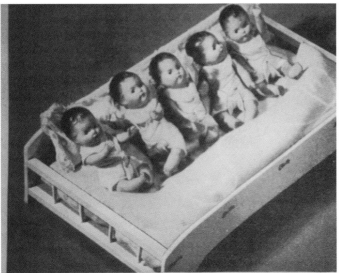

Left:
The John Plain catalog for 1937 offered a set of the 7" composition quints with curved baby legs along with a crib that featured the names of the girls printed on its side. *From the collection of Marge Meisinger.*

8" composition Alexander Dionne Quint straight leg toddler, circa 1935-38. She has painted eyes, closed mouth, and molded hair. Marked on back and neck: "Alexander." She is wearing an untagged blue cotton outfit, Cecile name pin, and snap shoes ($175+ with clothing wear). *Doll and photograph from the collection of Carol Stover.*

11" composition Alexander Dionne Quints straight leg toddlers, circa 1936-38. They have sleep eyes, closed mouths, molded hair, and are fully jointed. They are marked "Alexander" on their heads and "Madame Alexander" on their backs. They are dressed in their original organdy dresses and bonnets, but the ribbons on their bonnets and their shoes have been replaced. The pins may not be original (set $1,500+).

8" Alexander composition toddler Dionne Quints, circa 1937-38. The dolls have painted eyes, closed mouths, mohair wigs, and are fully jointed. They wear their original clothing consisting of organdy pastel dresses with matching bonnets and center snap shoes. Two are missing their bonnets. The dolls are marked "Dionne/Alexander" on their heads and "Alexander" on their backs (set $1,500+). *Dolls from the collection of Marge Meisinger. Photograph by Carol Stover.*

Personality Dolls 159

Left:
12" composition Alexander Dionne Quint straight leg toddler, circa 1937-38. She has sleep eyes, closed mouth, and molded hair. She is all original wearing a tagged yellow (Annette) cotton dress and bonnet. She is marked on the head "Alexander" and on her back, "Madame/Alexander." ($350+). *Doll and photograph from the collection of Carol Stover.*

Right:
16" Alexander toddler Dionne Quint, circa 1936-38. She has sleep eyes, closed mouth, and a human hair wig. She wears Yvonne's pink organdy dress with matching bonnet and pin, leatherette shoes, and is one of five in the owner's collection ($425+). *Doll from the collection of Marge Meisinger. Photograph by Carol Stover.*

The Sears Christmas catalog for 1936 featured several different sizes of Dionne Quint dolls on its cover. Babies with molded hair came in 7-1/2" and 11-1/2" sizes for 89 cents and $1.85 each. A set of the 7-1/2" babies with their bed was priced at $4.49. Larger wigged toddler dolls came in sizes of 11-1/2" and 14-1/2" inches. They cost $2.25 and $3.79 each. *From the collection of Linda Boltrek.*

Set of 16" Alexander composition Dionne Quints, circa 1936-38. The dolls have sleep eyes, closed mouths, and real hair wigs. They are all original with their pins and are wearing dotted swiss dresses and matching bonnets (not enough examples to determine a price.) *Dolls from the collection of Lois Jakubowski. Photograph by Robert Jakubowski.*

160 Personality Dolls

14" Alexander composition Dionne Quint, circa 1937. This appears to be a very similar doll to the models featured in the Sears Christmas catalog for that year. She has sleep eyes, closed mouth, and a real hair wig. She is all original in near mint condition, complete with her picture tag ($500+). *Doll from the collection of Lois Jakubowski. Photograph by Robert Jakubowski.*

Sears again featured Dionne Quint dolls in their 1937 Christmas catalog. Since the quints were growing up, the new dolls did not look like toddlers but were modeled more like the little girl dolls of the era. They came in sizes of 13", 17", and 20" and were priced at $2.98, $3.98, and $4.98 each. The dolls had sleep eyes, wigs, composition heads, arms, legs, and cotton stuffed bodies with cry voices. *From the collection of Marge Meisinger.*

Miscellaneous

Amosandra

Amosandra was the baby daughter of Amos and Ruby, characters from the "Amos 'n' Andy" radio program. The program first aired on March 19, 1928 on WMAQ in Chicago. The main black characters were played by white men, Freeman Gosden and Charles Correll. The show became one of the most popular and longest running radio programs ever. In the 1950s, the show changed format and became the "Amos 'n' Andy Music Hall." When the show began on television in 1951, black actors portrayed the characters.

The 1949 10" Amosandra rubber doll was a drink and wet model with painted features and jointed arms and legs.

10" rubber Amosandra doll made by the Sun Rubber Co., circa 1949. She has painted eyes, open mouth, molded hair, and is fully jointed. She was a "drink and wet" doll. The doll is marked "Amosandra/Columbia Broadcasting System Inc./Designed By/Ruth E. Newton/Mfg. By The Sun Rubber Co./Barberton/USA Pat. 2118682/2160739." She represents Amos and Ruby's baby from the radio show "Amos 'n' Andy." The clothing has been replaced ($100-125).

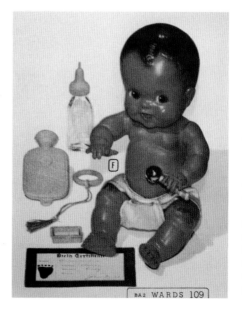

The 1949 Montgomery Ward Christmas catalog advertised the Amosandra rubber doll for $2.98. It was 10" tall, had painted eyes, jointed arms and legs, with a head that turned. The doll could drink, wet, and cry. *From the collection of Betty Nichols.*

Carol Ann Berry

Carol Ann Berry was the adopted daughter of the famous screen star, Wallace Berry. She appeared in several of her father's films, including *China Seas* in 1935.

The American Character doll firm marketed a doll that same year that was supposed to represent the young girl. The doll was actually a regular composition Sally model with an unusual hairstyle added. The braid across the top of the doll's head and her "Two-Some" costumes identified her as Carol Ann Berry. The dolls had sleep eyes, closed mouths, mohair wigs, and were fully jointed. They were produced in 13", 16", and 19" sizes. Each doll came with outfits that could be used in two ways, including a dress that, when removed, revealed a playsuit instead of underwear. Other models were dressed in a robe and pajamas but carried an extra dress, or in a coat and hat with a dress under the coat. Trunk sets with extra clothing were also available.

Fanny Brice

Fanny Brice was born Fanny Borach in 1891 in New York City. The star had a long and varied career. She headlined the *Ziegfeld Follies* for many years beginning with a small part in 1910. She first played her "Baby Snooks" character in the Follies. She introduced the character to a radio audience in 1938 and continued to do a radio show starring Baby Snooks until 1951. Fanny also made several films, including *The Great Ziegfeld*, *My Man*, and *Everybody Sing*. Her earlier career featured Fanny as a singer as well as a comedienne. One of the most memorable songs identified with Fanny was "My Man."

The Ideal company marketed a doll in the image of Baby Snooks in 1939 to tie in to the radio program's success. The doll was 12" tall with a composition head and hands, a wooden torso and feet, and flexy wire arms and legs. She had molded hair, painted features, and an open-closed mouth. The doll was designed by Joseph Kallus. The flexy wire arms and legs allowed the doll to hold various poses. A companion Mortimer Snerd doll was also made based on another well-known radio character (see Charlie McCarthy).

Modern audiences are familiar with Fanny from the two Barbra Streisand films *Funny Girl* (1968) and *Funny Lady* (1975), based on the life of Fanny Brice. Fanny died in 1951 after a career lasting over forty years.

American Character's Carol Ann Beery composition doll was advertised in the Sears 1936 Christmas catalog. She had sleep eyes, closed mouth, a mohair wig with a braid across the top of her head, and was fully jointed. She was dressed in a sailor dress and a matching playsuit was also included. A picture of the famous movie actor, Wallace Beery, along with his daughter Carol Ann came with the doll. *From the collection of Linda Boltrek.*

Right:
12" Ideal Baby Snooks (Fanny Brice), circa 1938-39. The head and hands are composition, the torso and feet are wood, and the arms and legs are made of a flexy wire. She has painted eyes, open mouth with painted teeth, and molded hair. The head is marked "Ideal Doll." She is all original. The doll was designed by Joseph Kallus and represented the character Fanny Brice played on her radio program. Pictured with the doll is the *Movie-Radio Guide* magazine from Nov. 21-27, 1942, which featured Fanny Brice in her Baby Snooks costume on its cover.

Below:
The Sears Christmas catalog for 1939 advertised the Fanny Brice doll, along with the matching Ideal Mortimer Snerd doll and the Ideal Walt Disney Pinocchio doll. Each doll was priced at only 95 cents. The Pinocchio doll was 11" tall while the other dolls were 12-1/2"-13" tall. *From the collection of Marge Meisinger.*

162 Personality Dolls

Deanna Durbin

Deanna Durbin was born in Winnipeg, Canada in 1922 as Edna Mae Durbin. She and her parents moved to Los Angeles, where it was discovered that she had an unusually remarkable singing voice for a child. In 1935, she was signed by M-G-M and made a short film with another newcomer to the industry, Judy Garland.

Since her career seemed stalled, Deanna was soon dropped by the studio. She then began singing on the Eddie Cantor radio show and became very popular. This led to a contract with Universal Pictures and her first film, *Three Smart Girls*, was released in 1936. The picture grossed two million dollars, and with the continuing success of Deanna's movies, her films were credited with making Universal Pictures a major movie studio.

Because Deanna had become such a popular star, Ideal began making a composition doll in her image in 1938. The dolls had sleep eyes, open mouths with teeth, human hair wigs, and were fully jointed. The dolls were dressed in teen-age clothing. They were marked on the back of the head "Deanna Durbin" and were made in a variety of sizes, including 14-1/2", 17-1/2", 21", and 25". The same body was also used as the body for the teen Judy Garland. The Deanna Durbin doll was also dressed as Gulliver from the 1939 cartoon film *Gulliver's Travels*. She had nothing to do with the movie, it was just a case of using the mold of a doll already being made and marketing it as something else.

Deanna Durbin continued to be a star through the World War II years. She made a successful transition to adult roles but her popularity began to drop during those years. She retired from pictures after making her last movie for Universal in 1949—she was only twenty-seven years old. She married the French director Charles-Henri David in 1950 and moved to France.

Some of Deanna Durbin's best films include: *One Hundred Men and a Girl* (1937), *Mad About Music* (1938), *That Certain Age* (1938), *Christmas Holiday* (1944), and *Lady On A Train* (1945). All of her movies were made for Universal.

18" Ideal composition Deanna Durbin, circa 1939-40. She has sleep eyes, open mouth with teeth, real hair wig, and is fully jointed. The Durbin dolls were marked "DEANNA DURBIN/IDEAL DOLL" on their heads. She is all original, in beautiful condition, complete with her pin ($750+). *Doll from the collection of Lois Jakubowski. Photograph by Robert Jakubowski.*

Right:
The John Plain catalog for 1940 pictured an Ideal 18" composition Deanna Durbin doll dressed in a white organdy party dress trimmed with blue checked taffeta. An autographed picture of Deanna was included with the doll. *From the collection of Marge Meisinger.*

The Speigel catalog for 1939 featured an unusual Ideal doll advertisement in color. Shirley Temple dolls in sizes of 16", 18", and 22" were available as well as Judy Garland "Dorothy" dolls in 15" and 18". The Deanna Durbin dolls offered were 15" and 21" tall. The dolls ranged in price from $2.98 to $5.98 each. *From the collection of Marge Meisinger.*

Left:
21" Ideal composition Deanna Durbin, circa 1938-40. She has sleep eyes, open mouth with teeth, human hair wig, and is fully jointed. She is all original wearing her tagged dress. The Deanna dolls were so popular that the dolls were marketed in many different outfits, including both long and short dresses ($650+). *Doll from the collection of Lois Jakubowski. Photograph by Robert Jakubowski.*

Personality Dolls

21" Ideal composition Deanna Durbin, circa 1938-40. She wears a long gown with a pink taffeta skirt and gold brushed satin top with green sandals. She is all original, including her pin ($700+). *Doll from the collection of Marge Meisinger. Photograph by Carol Stover.*

21" Ideal composition costumed Gulliver using a Deanna Durbin doll, circa 1939. The doll represented the main character in the cartoon film, *Gulliver's Travels* from 1939. Deanna Durbin had nothing to do with the movie, but Ideal used a doll already on the market to make the new character. This model had painted eyes and a black mohair wig. The doll is all original and much harder to find than a regular Deanna Durbin doll ($850+). *Doll from the collection of Lois Jakubowski. Photograph by Robert Jakubowski.*

W.C. Fields

William Claude Dunkenfield was born in Philadelphia in 1879. He ran away from home at the age of eleven and became a street child. He learned to juggle and began to entertain in vaudeville shows by his late teens. In 1905, he appeared in his first Broadway play and by 1915 he was starring in silent films. With the coming of sound, he became a major movie star. His most popular movies included *Tillie and Gus* (1933), *David Copperfield* (1935), and *My Little Chickadee* (1939) with Mae West. He was also active in radio during the 1930s, making many guest appearances. He died in 1946.

A composition ventriloquist doll was made to represent W.C. Fields in 1938 by the Effanbee company. The doll was 17-1/2" tall with a composition head, hands, and shoes and a cloth body. The doll had a ventriloquist mouth with painted teeth, painted eyes, and molded hair. He was dressed in a top hat and suit.

Left:
21" Ideal composition Deanna Durbin, circa 1938-40. She is all original, wearing a long gown with a plaid taffeta skirt, matching hat and umbrella, white bodice, and lace gloves ($700+). *Doll from the collection of Lois Jakubowski. Photograph by Robert Jakubowski.*

Right:
The N. Shure Co. catalog from 1938 advertised ventriloquist dolls in the images of the movie actor, W.C. Fields, and the radio star, Charlie McCarthy. The actor and the "Dummy" had been having a feud on the Edgar Bergen radio show for many months so the copy on these ads stated that W.C. Fields could "talk back" to Charlie. Both dolls were made in the ventriloquist dummy style and Fields came in a 20" size while Charlie could be purchased in 14", 22", or 18" sizes. Both dolls had composition heads, painted features and molded hair. The bodies were stuffed. *From the collection of Marge Meisinger.*

Judy Garland

Judy Garland was born in Grand Rapids, Minnesota on June 10, 1922 as Frances Ethel Gumm. Her mother was not content to remain a small-town wife and mother, so she staged an act with her three young daughters.

Although the Gumm sisters were moderately successful in the Grand Rapids area, there was limited opportunity for growth. Mrs. Gumm talked her husband into moving to California so the girls would have a better chance for success. The sisters continued to perform in California wherever they could obtain bookings. When Judy was thirteen, she was called to M-G-M for an audition and their luck changed. Judy sang "Zing Went the Strings of My Heart" and was given a contract by the studio.

Judy's movie career began with a small short called *Every Sunday*, which she made with another newcomer, Deanna Durbin. Her big break came when she received the part of Dorothy in *The Wizard of Oz*. With the success of the Oz movie, Judy's career gained momentum and she began a series of eight films with Mickey Rooney. Some of the movies were part of the Andy Hardy series while others were Busby Berkeley musical extravaganzas. Some of the best were *Babes in Arms*, *Strike Up the Band*, *Babes on Broadway*, and *Girl Crazy*.

When Judy became successful in the movies, the Ideal Co. began production of two dolls made in her image. The first doll represented her in her role as Dorothy in *The Wizard of Oz*. The composition doll was available in 1939 and was designed by famous doll designer, Bernard Lipfert. The doll was dressed in a copy of Judy's "Oz" dress and had sleep eyes, an open mouth with teeth, and a human hair wig. The dolls were made in 14", 16", and 18" sizes. A more grown-up teen Judy doll was marketed in 1941. This composition doll was dressed in a copy of a costume from the M-G-M film, *Strike Up the Band*. The dolls came in 18" and 21" sizes and had sleep eyes, open mouths with teeth, and real hair wigs.

Judy's career at M-G-M continued in high gear during the war years as she made many of the studio's most successful musicals. These included *Ziegfeld Girl*, *For Me and My Gal*, and *Meet Me In St. Louis*. After the war, Judy began to experience health problems and her many collapses and hospital stays interfered with her successful career. She still managed to make a number of fine films for M-G-M, however, including *The Harvey Girls*, *The Pirate*, *Easter Parade*, *In the Good Old Summertime*, and *Summer Stock*.

In 1950, Judy was fired from M-G-M because of her health problems. She was only twenty-eight years old. She pulled herself together and turned to concerts. She did a smash performance at the London Palladium, and then played the Palace Theatre in New York in 1951. With the new interest in Judy as a performer, her husband Sid Luft became her producer and she starred in a successful remake of the film, *A Star Is Born*.

For the rest of her life, Judy concentrated mostly on concert and cabaret work. She did branch into television in 1963 when she did a weekly television variety show. It was not a success and was canceled after twenty-six episodes. Judy Garland died on June 22, 1969, at the age of forty-seven, from an overdose of sleeping pills.

16" Ideal composition Judy Garland "Dorothy" doll from *The Wizard of Oz*, circa 1939. She has sleep eyes, open mouth with teeth, a real hair wig, and is fully jointed. She is marked "IDEAL DOLL/MADE IN USA" on her head. She is all original including her black shoes with the flaps as pictured in the Sears ad from 1939. The doll is shown with the original "Over The Rainbow" sheet music from *The Wizard of Oz* M-G-M movie. The song is by E.Y. Harburg and Harold Arlen. Leo Feist Inc. published this sheet music in 1939 (doll $1,000+, music $35+).

Several composition personality dolls were advertised in the Sears Christmas catalog for 1939. Included were Sonja Henie, Cinderella, Deanna Durbin, and Judy Garland. The Sonja dolls came in sizes of 14" and 18" and were priced at $2.79 and $4.69. Deanna was advertised for $2.98 in a 14" size and $4.98 for a 21" doll. Cinderella only came in a 13-1/2" size and was priced at $2.98. The Judy Garland doll was listed in a 14" size for $2.79 and an 18" size costing $4.79. *From the collection of Marge Meisinger.*

18" Ideal composition Judy Garland "Dorothy" doll, circa 1939 ($1,200+). *Doll from the collection of Lois Jakubowski. Photograph by Robert Jakubowski.*

Rita Hayworth

Rita Hayworth was christened Margarita Carmen Cansino when she was born in 1918 in New York City. Her parents were both dancers, so Marguarita began taking dancing lessons about as soon as she could walk. During the Depression, when Margarita was fourteen, she joined her father as his dancing partner for shows in Mexico.

It was in Mexico that Margarita was seen by Winfield Sheehen, from Fox studios, and was given a Fox movie contract. Her first film, *Dante's Inferno*, was released in 1935. She made several more movies for Fox but received little notice. When Fox merged with Twentieth Century, Rita was out of a job.

With the guidance of Edward Judson (whom Rita later married), she secured a contract from Columbia in 1937 and her name was changed to Rita Hayworth. Although she made several minor films, Rita was not really noticed until she received a role in the picture, *Only Angels have Wings*, in 1939. With this new interest in Rita, Columbia starred her in a musical called *Music In My Heart* in 1940 and loaned her to Warner's for *Strawberry Blonde* in 1941. Twentieth Century-Fox also borrowed Rita in 1941 to make *Blood and Sand*. Rita Hayworth was suddenly a star. Some of her most famous pictures during the war years include *You'll Never Get Rich*, *You Were Never Lovelier*, *My Gal Sal* (Twentieth Century-Fox), *Cover Girl*, and *Tonight and Every Night*.

Unlike most of the pin-up stars of the early 1940s, Rita's career did not slow down at the end of World War II. One of her most remembered pictures, *Gilda*, was not made until 1946. Rita continued her successes for Columbia with *Down to Earth* in 1947, *The Loves of Carmen* in 1948, *Affair in Trinidad* in 1952, and

The 1941 Spiegel catalog featured both a teen Deanna Durbin and a teen Judy Garland doll for sale. The Deanna dolls came in 15" and 21" sizes and were priced at $4.45 and $6.45 each. The Judy Garland dolls were 18" and 21" tall and cost $4.45 and $6.45 each. Both composition dolls wore long gowns and were made by Ideal. Judy's dress was based on a costume she wore in the M-G-M film *Strike Up the Band*. From the collection of Marge Meisinger.

18" tall composition Ideal "Teen" Judy Garland, circa 1941. She has sleep eyes, open mouth with teeth, real hair wig, and is fully jointed. She is very faintly marked "Deanna Durbin" on her back but Ideal was known for using Shirley Temple and Deanna Durbin bodies on other dolls. She is wearing the long dress from *Strike Up the Band* that was shown on the "Teen" Judy Garland in the 1941 Speigel catalog (original with some wear $450+).

15" composition Rita Hayworth Carmen doll, circa 1948. She was made by Uneeda Doll Co. The doll has sleep eyes, closed mouth, mohair wig, and is fully jointed. She is unmarked but originally came with a wrist tag identifying her as a "Carmen Doll inspired by Rita Hayworth's Carmen in the movie The Loves of Carmen." She is all original wearing gold shoes and a red taffeta dress with a black mesh overskirt and mantilla, both trimmed with red flowers. The bodice of the dress is accented with gold rick-rack. Pictured with the doll is a set of uncut "CARMEN PAPER DOLLS AS INSPIRED BY THE PERFORMANCE OF RITA HAYWORTH AS CARMEN IN /THE LOVES OF CARMEN/A COLUMBIA TECHNICOLOR PICTURE/A BECKWORTH CORP. PRODUCTION." #1592 Copyright 1948 by the Saalfield Publishing Co. By Permission of Walter I. Gould & Co. (doll $400+, paper dolls $50+).

166 Personality Dolls

Salome in 1953. She also teamed with her husband, Orson Wells, and starred in *The Lady From Shanghai* in 1948. Although Rita continued to make films until 1972, she was no longer the star she had been. Her best movie from this period was *Separate Tables*, made by United Artists in 1958.

Rita's personal life continued to make headlines with marriages and divorces. Additional husbands included Prince Aly Khan, singer, Dick Haymes, and screenwriter James Hill.

Rita Hayworth died, after a long illness, in May 1987 from Alzheimer's disease.

Sonja Henie

Sonja Henie made her way to the Hollywood screen in a very unique way—she traveled the road to fame and fortune on ice skates.

Born in Oslo, Norway in 1912, Sonja began skating at the age of six and by fourteen was declared the ice skating champion of Norway. In 1927, she was awarded the world skating championship, a title she held for ten years. In 1928, she won her first Olympic gold medal in figure skating and went on to win again in 1932 and in 1936.

After the 1936 Olympics, Sonja and her father planned a skating tour in the United States. Darryl Zanuck from Twentieth Century-Fox saw the show and signed Sonja to a five year contract with his studio. Sonja's first film, *One In a Million*, was released in 1936 and became an immediate success. She continued to make skating pictures for Fox for several years. Her best movies include *Happy Landing*, *Thin Ice*, *My Lucky Star*, *Second Fiddle*, *Sun Valley Serenade*, and *Iceland*.

Because of Sonja's popularity in the movies, the Alexander firm began producing dolls in the star's image in 1939. They came in sizes of 13-1/2" (with a swivel waist), 14", 17-1/2", and 21". The dolls were made of composition and were marked on the head "Madame Alexander/Sonja Henie." They had sleep eyes, open mouths with teeth, and either mohair or human hair wigs.

Although Sonja was a successful Hollywood movie star, she still wanted to do more to promote ice skating. In 1938, she and a partner, Arthur Wirtz, began the *Hollywood Ice Revue*. The show was to become an annual touring event. Sonja herself skated in the revue until 1952.

Sonja made her last film, *The Countess of Monte Carlo*, for Universal in 1948. The novelty of a skating movie star had worn off so Sonja retired from pictures. During her career, she had invented the ice musical and made eleven films in twelve years, which grossed a total of twenty-five million dollars. Sonja Henie died from leukemia in 1969 at the age of fifty-seven.

14" composition Alexander Sonja Henie, circa 1939-41. She is all original wearing a very rare skating outfit and is in excellent condition. She has a mohair wig ($550+). *Doll and photograph from the collection of Nancy Roeder.*

Right:
14" composition Alexander Sonja Henie, circa 1939, with the Wendy Ann swivel waist body, sleep eyes, open mouth, and real hair wig. She wears her original pink taffeta costume with maribou trim and white skates. The pin was sold separately ($400+ with paint flecking). *Doll from the collection of Marge Meisinger. Photograph by Carol Stover.*

Left:
14" composition Alexander Sonja Henie, circa 1939-41. She has sleep eyes, open mouth, mohair wig, and is fully jointed. She is marked on her head "Madame Alexander/Sonja/Henie." She wears her original clothing but her skates have been replaced and the flowers in her hair have been added. Pictured with her are "Sonja Henie Paper Dolls" #3475. Copyright U.S.A. 1939 by Merrill Publishing Co. Chicago, Illinois (doll $350 with replaced skates, uncut paper doll book $175-200).

Personality Dolls 167

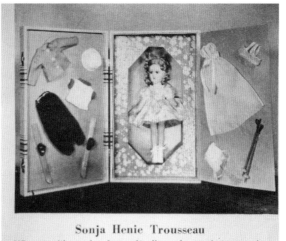

The Carson Pirie Scott & Co. catalog carried a very nice Sonja Henie Trousseau set in 1941. The 15" doll came with a dress, underwear, skates, ski outfit, skis, poles, nightgown, bra, and girdle. *From the collection of Marge Meisinger.*

The N. Shure Co. advertised Sonja Henie dolls in their 1941 catalog. The dolls came in sizes of 15", 18-1/2", and 22" in three different outfits. These dolls all had human hair wigs. *From the collection of Marge Meisinger.*

Alexander Sonja Henie Trousseau set as pictured in the 1941 Carson catalog. The set is complete and in its original trunk. The later pink Alexander accessory box did not come with the set and the purse may also be a later addition. What is surprising is the inclusion of a bra and girdle. Those items were not usually part of a doll's wardrobe until the 1950s (not enough examples to determine a price). *Doll from the collection of Lois Jakubowski. Photograph by Robert Jakubowski.*

Left:
14" composition Alexander Sonja Henie wearing a costume similar to those in the N. Shure Co. catalog. She also has the human hair wig. She wears her original yellow taffeta costume and gold skates. The pin was sold separately ($450-500). *Doll from the collection of Marge Meisinger. Photograph by Carol Stover.*

18" composition Alexander Sonja Henie, circa 1940-41. She is all original in excellent condition. She is wearing a taffeta and net skating costume, matching head piece, and gold skates ($800). *Doll from the collection of Lois Jakubowski. Photograph by Robert Jakubowski.*

Sandra Henville (Baby Sandy)

Sandra Lee Henville had the distinction of being the youngest movie "star" during the heyday of her film career from 1939-41. She was born in Los Angeles, California in 1938 and went to work at Universal Pictures when she was only a year old. She began her acting career by playing the roles of boy babies in her first two films, *East Side of Heaven* and *Little Accident*, both made for Universal in 1939. The studio called her Baby Sandy in order to disguise the fact that she was a girl.

Other Baby Sandy movies include *Unexpected Father*, *Sandy is a Lady*, *Sandy Gets Her Man*, *Melody Lane*, *Bachelor Daddy*, and *Johnny Doughboy*.

Sandra Lee Henville never made another film and she retired to private life, eventually to become a housewife and the mother of three sons. She also had a long career as a legal secretary.

Books, dolls, paper dolls, cups, and other toys were produced in the Baby Sandy image. The dolls are the most collectible of these products. Ralph A. Freundlich Inc. of New York produced the dolls in four sizes, including 8", 12", 16" and 20". The smaller dolls had painted eyes while the larger dolls had sleep eyes and some had wigs. The dolls were marked on the back of their necks "BABY SANDY."

12" composition Freundlich Baby Sandy, circa 1939-42. The doll has painted eyes but the 16" and 20" dolls had sleep eyes. All of the dolls were marked "Baby Sandy" on their heads. She is all original, including her pin, and is wearing an organdy dress ($275-300). *Doll from the collection of Marge Meisinger. Photograph by Carol Stover.*

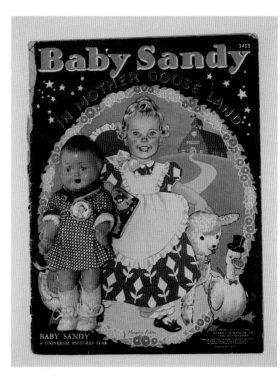

8" composition Freundlich Baby Sandy, circa 1939-42. She has painted eyes, closed mouth, molded hair, and is fully jointed. She is marked on the back of her head "Baby Sandy." The doll is all original including her Baby Sandy Button. It reads "NEW UNIVERSAL STAR/GENUINE BABY SANDY DOLL." Pictured with the doll is the book *Baby Sandy in Mother Goose Land* (#3425) published by the Merrill Publishing Co. in 1940 (doll $175-225, book with wear $25).

Vivien Leigh

Vivien Leigh was born in India in 1913 with the name Vivien Mary Hartley. Her first movie was made in England in 1934. She became a star in 1936 when she played opposite Laurence Olivier in *Fire Over England*. She received the coveted role of Scarlett O'Hara in *Gone With the Wind* when she was in the United States with Laurence Olivier while he was making a film. She won an Academy Award for her performance and the Academy awarded her another for her great role in *A Streetcar Named Desire* in 1951. She also played in several more films and in many stage productions including those she did with her husband, Olivier. She was plagued with poor health and died at the age of fifty-two in 1967. She and Olivier had divorced five years earlier.

Although the Alexander Scarlett O'Hara dolls have been marketed for many years, collectors like to think that, at least, the early dolls represent Vivien Leigh in her "GWTW" role. The Alexander company never advertised the dolls as Vivien Leigh dolls, however. The first Alexander Scarlett dolls were marketed in 1940-41. They eventually came in sizes of 11", 14", 16", 18", and 21". The dolls had sleep eyes, closed mouths, real hair black wigs, and were fully jointed. The tagged clothes and boxes identified the Scarlett dolls.

11" composition Alexander Scarlett O'Hara, circa 1940-41. She is all original, complete with her box, dress tag, and sticker on her dress reading "Scarlett." (MIB $1,000+). *Doll and photograph from the collection of Nancy Roeder.*

11" composition Alexander Scarlett O'Hara, circa 1940-41. She has sleep eyes, closed mouth, black mohair wig and is fully jointed. The tag on her original dress reads "Scarlett O'Hara/by Madame Alexander N.Y./All rights Reserved." She is all original except for the flower in her hair which has been replaced. She is pictured with a *GONE WITH THE WIND COOK BOOK* picturing Vivien Leigh in a scene from the famous film. The book was a premium with the purchase of Pebeco toothpaste. "Inspired by the picture "GONE WITH THE WIND"/A SELZNICK INTERNATIONAL PICTURE./Produced by David O. Selznick/A Metro-Goldwyn-Mayer Release." (doll with some wear on dress and replaced flower $300, book $35).

11" composition Alexander Scarlett, circa 1940-41. She has green eyes, a human hair wig, and wears a cotton print dress with white cotton slip and a straw hat. She is all original with her box ($950+). *Doll from the collection of Marge Meisinger. Photograph by Carol Stover.*

Personality Dolls

14" composition Alexander Scarlett O'Hara, circa 1940-41. She is all original wearing a tagged Scarlett O'Hara dress and coat and hat. Her dress also has a gold sticker on the front reading "Scarlett." ($1,000+). *Doll and photograph from the collection of Nancy Roeder.*

The 1941 Spiegel catalog advertised an Alexander Scarlett O'Hara doll in a 14-1/2" size for $3.98. The doll wore a hoop petticoat and pantaloons as did most of the Scarlett dolls. The copy reads "Just like the heroine of 'Gone With the Wind.'" There is no mention of Vivien Leigh, but collectors like to assume the dolls represent the lovely actress who played the Scarlett role. *From the collection of Marge Meisinger.*

14" composition Alexander Scarlett, circa 1940-41. She is all original including her hoop petticoat and pantaloons. She is wearing a tagged yellow taffeta dress, matching hat, and black jacket ($950+). *Doll from the collection of Lois Jakubowski. Photograph by Robert Jakubowski.*

18" composition Alexander Scarlett O'Hara, circa 1940-41. She has green sleep eyes, closed mouth, and a human hair wig. She is all original wearing a checked gown with hoop skirt, pantalettes, and green leatherette shoes. She has a wrist tag personally signed by Madame Alexander. The tag reads "Madame Alexander/Scarlett O'Hara Doll" along with a drawing of Scarlett wearing her famous white dress (not enough examples to determine a price.) *Doll from the collection of Marge Meisinger. Photograph by Carol Stover.*

Lone Ranger and Tonto

In 1938, the Dollcraft Novelty Co. produced dolls representing the Lone Ranger and Tonto. The fictional characters had gained fame in both films and on the radio. Created by George W. Trendle and Fran Striker, the characters were brought to life on a radio series in 1933, followed by several films, then a television series beginning in 1949. They were also featured in a comic strip in 1955. Jay Silverheels played Tonto in two of the films and the television series while Clayton Moore played the part of the Lone Ranger in movies and most of the television programs.

The dolls had composition heads and hands, cloth bodies, and molded shoes. They had painted features and hair and were 20-1/2" tall. The pair of dolls are very hard to find and are very collectible.

Douglas MacArthur

Douglas MacArthur was a very famous general who gained fame during World War II. He was born in 1880 and graduated from West Point. During World War I, MacArthur commanded the 42nd Division. He continued to make his mark in the military and retired in 1937. When the United States became involved in World War II, MacArthur was recalled to duty. He became the Supreme Allied Commander of the South West Pacific during the war and after Japan surrendered, he became commander of the occupational forces in Japan.

16" composition General Douglas MacArthur doll made by the Ralph Freundlich firm, circa 1942. He has painted features, molded hat and hair, and is jointed at the shoulders and hips. His right arm is bent so he is able to salute. He is all original in his uniform. Shown with the doll is the *Life* magazine for December 8, 1941, which featured the General on its cover ($250 with some wear to hat).

Ralph A. Freundlich Inc. marketed a doll in the General's image circa 1942. The 18" composition doll had a molded hat and painted features and was dressed in a uniform based on the one worn by MacArthur. His right arm was curved so that it could be used to salute. Only the original paper tag identified the doll as being General MacArthur.

The General was Commander in Chief of the United Nations forces in Korea in 1950-51 but was fired by President Harry Truman when they had a conflict about the war. He died in 1964.

Charlie McCarthy (and Edgar Bergen)

In 1936, Edgar Bergen and his ventriloquist dummy, Charlie McCarthy, made a guest appearance on the popular Rudy Vallee radio show. Their act was such a hit, they aired for three more months.

In 1937, Bergen started his own show sponsored by Chase and Sanborn. The program was the number one rated radio show in the United States for two and a half years. Even as late as 1945, it was in fifth place after having aired for eight years. The supporting cast for the show was also excellent. Don Ameche was the master of ceremonies, the singer was Nelson Eddy, and Ray Noble led the orchestra. The comedian W.C. Fields was a frequent guest on the show and from 1937-39, the "on the air" feud between Charlie and Fields provided material for many of the best programs from the series. Bergen and McCarthy were regulars on radio programs until 1956 when they moved to television.

Perhaps because Charlie McCarthy (along with another Bergen ventriloquist dummy-Mortimer Snerd) couldn't be seen by the radio audience, they seemed like real people to the show's fans. In 1938, Effanbee marketed dolls to represent the famous Charlie. These dolls proved to be good sellers and other companies continued to produce Charlie ventriloquist dolls from time to time for decades. The Effanbee products included a variety of styles and sizes. The dolls came in 15", 17", and 20" sizes. They were made with composition heads and hands, and molded shoes and socks. The bodies, arms, and legs were cloth. The dolls had molded hair and painted features. Each doll came with the famous Charlie monocle. The mouth could be moved so that the doll could be used as a ventriloquist dummy. They were marked "Edgar Bergen's/ Charlie McCarthy/an Effanbee Product."

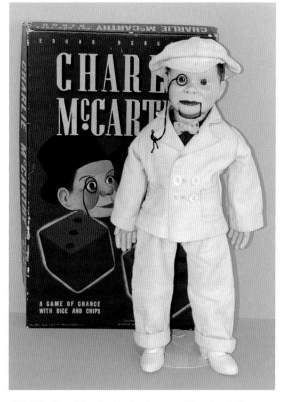

15" Effanbee Charlie McCarthy ventriloquist doll, circa 1938-39. He has a composition head with a mouth that can be moved, composition hands and feet, and a cloth body and limbs. Charlie has molded hair and painted features. His head is marked: "Edgar Bergen/ Charlie McCarthy/An Effanbee Product." He is missing his pin but otherwise is all original complete with his monocle. He is pictured with a "Charlie McCarthy Put and Take Bingo" game. It was © 1938 by McCarthy Incorporated No. 2913, Whitman Publishing Co. (doll $400, game $40.)

Personality Dolls

The Ideal company marketed a 12" Mortimer Snerd doll to tie in to the famous radio show in 1939. He was one of their Flexible dolls and was marketed at the same time as the Baby Snooks doll. He had a composition head and hands, a wood torso and feet, and wire arms and legs. His hair was molded and he had painted features. He was marked "Ideal Doll."

Carmen Miranda

Carmen Miranda was born in Marco de Canavezes, Portugal in 1909. Her real name was Maria do Carmo Miranda da Cunha. She and her parents soon moved from Portugal and settled in Brazil, where she became both a movie and singing star.

After her success in Brazil, Miranda came to the United States to do a Broadway musical. She was signed for movies by Twentieth Century-Fox and in 1940 made her first film, *Down Argentine Way*. Miranda became known for her trademark turban hats, which were piled high with fruit. Her costumes also included lots of jewelry and high heeled platform shoes.

In the early 1940s, during the height of Carmen's movie career, the Alexander company produced several sizes of composition dolls called "Carmen." The dolls were dressed with fancy turban hats and Miranda-like costumes but the company did not list the dolls as Carmen Miranda dolls. There is no doubt, however, that the dolls were made to tie in to the star's popularity. The smaller 7" and 9" dolls came with painted eyes while the larger 11", 15", 18" and 22" dolls were made with sleep eyes. The dolls had closed mouths and were fully jointed. Only the original dress tags identify the dolls as "Carmen."

Carmen Miranda made ten films for Fox and then did four more movies for other studios before her motion picture career ended. She died of a heart attack in 1955. Some of her best films were *Weekend in Havana*, *Springtime in the Rockies*, *The Gang's All Here*, and *Four Jills in a Jeep*.

The John Plain catalog for 1939 carried an advertisement for two sizes of the Charlie McCarthy ventriloquist dolls. A booklet on how to be a ventriloquist came with each doll. The dolls were 15" and 20" tall. *From the collection of Marge Meisinger.*

Left:
12" Ideal "flexy" Mortimer Snerd, circa 1939. He was one of Edgar Bergen's ventriloquist "dummies." The doll has a composition head and hands, wood body and feet, and wire posable limbs. His hair and features are painted. He is marked "Ideal Doll/ Made in USA." He wears his original clothing. Also pictured is a wind-up tin Mortimer Snerd figure made by Louis Marx and Co. in 1939. On the back it says, "Meet my friend Mortimer Snerd Signed Charlie McCarthy." It is marked "© 1939 McCarthy Inc." (doll $225, wind-up $200-225).

14" composition Alexander Carmen, circa 1942. Although the company did not claim that the doll represented Carmen Miranda, there is no doubt of its tie-in to the popular movie star. She has sleep eyes, closed mouth, black mohair wig, and is fully jointed. She wears her original dress but her turban and shoes and socks have been replaced. The tag on her dress reads "Carmen/ Madame Alexander, N.Y. U.S.A./All Rights Reserved." On the back of head is marked "MD. Alexander." (doll not entirely original $350+).

7" and 9" composition Alexander Carmen dolls, circa 1942. The dolls have painted eyes, closed mouths, and black mohair wigs. They are the Tiny Betty and Little Betty dolls dressed in Carmen costumes and both are all original ($200-300 each). *Dolls from the collection of Marge Meisinger. Photograph by Carol Stover.*

Margaret O'Brien

Margaret O'Brien was born in Los Angeles in 1937 as Angela Maxine O'Brien. She began her career as a model and then worked her way into films. Margaret played a bit part in the 1941 M-G-M movie *Babes On Broadway,* but it was not until she received her first good role in the M-G-M film *Journey For Margaret* in 1942 that she earned well-deserved attention. She adopted her character's name as her own and was, thereafter, known as Margaret O'Brien.

With the success of the 1942 movie, Margaret's popularity skyrocketed. Although her career spanned a period of nineteen years and included a total of twenty-one pictures, over one-third of her movie output occurred during 1943 and 1944.

Some of her most successful films from this period include *Thousands Cheer, Lost Angel, The Canterville Ghost, Music For Millions, Meet Me in St. Louis,* and *Our Vines Have Tender Grapes.*

The Madame Alexander Doll Co. marketed Margaret O'Brien dolls beginning in 1946. The composition dolls came in sizes of 14", 18", and 21". The dolls had sleep eyes, closed mouths, mohair wigs in braids, and were fully jointed. By 1948, the dolls were being made of hard plastic.

Margaret continued to make films for M-G-M during the late 1940s with one big hit, *Little Women,* released in 1949. Her last picture was *Heller in Pink Tights,* made in 1960.

After the O'Brien film career was finished, the young actress kept busy in television and stock productions.

14" composition Alexander Margaret O'Brien, circa 1946-47. She has sleep eyes, closed mouth, mohair wig, and is fully jointed. She is unmarked. The doll is all original, wearing a pink organdy and ivory lace dress, taffeta slip and panty, pink leatherette shoes, and white socks. On her head is a straw hat. She is pictured with an uncut set of Margaret O'Brien paper dolls #970, published by the Whitman Publishing Co. in 1944 (doll $700-900, paper dolls uncut $175-200).

Right:
14" composition Alexander Margaret O'Brien, circa 1946-47. She is all original and wears a jumper dress and a straw hat ($700-900). *Doll and photograph from the collection of Nancy Roeder.*

18" composition Alexander Margaret O'Brien, circa 1946-47. She wears a similar jumper dress to the 14" doll along with a straw hat. She is all original including her wrist tag, which reads "A/Madame Alexander/Doll." ($1,100-1,300). *Doll from the collection of Marge Meisinger. Photograph by Carol Stover.*

174 Personality Dolls

Left:
18" composition Alexander Margaret O'Brien, circa 1946-47. She is all original wearing a checked cotton dress with embroidered nylon trim and a straw hat. The dolls were advertised in the Sears 1947 Christmas catalog in sizes of 14-1/2", 18", and 21", priced at $7.49, $9.98, and $12.98 ($1,100-1,300). *Doll from the collection of Marge Meisinger. Photograph by Carol Stover.*

Princesses (Elizabeth and Margaret Rose)

Princess Elizabeth (later to become Queen Elizabeth II) was born in London in 1926 to the Duke and Duchess of York. Her sister, Margaret Rose, was born in 1930. Their father became King George VI in 1936 when Edward VII abdicated in order to marry "the woman he loved." Elizabeth became heir to the throne at the age of ten.

The attractive little girls gained the world spotlight when their father became King, and many products were marketed in their image. Included were paper dolls and dolls. The Madame Alexander firm produced composition dolls of the girls and the Princess Elizabeth dolls were first made in 1937. These dolls had closed mouths while dolls with open mouths were marketed from 1938 to 1941. The dolls were made in a variety of sizes including 13"-14", 15"-16", 17"-18", 20", 24", and 27". Composition Alexander Princess Margaret dolls were advertised in the 1947 Montgomery Ward Christmas catalog in sizes of 15", 18", and 21". The dolls had mohair wigs, sleep eyes, and closed mouths. By 1949, the Margaret dolls were being made of hard plastic.

In 1947, Elizabeth married Philip Mountbatten, formerly Prince Philip of Greece. Their first child, Charles, was born in 1948, followed by Ann in 1950, Andrew in 1960, and Edward in 1964. Elizabeth became Queen of England in 1952 at the death of her father. She was twenty-five years old.

Princess Margaret married Anthony Armstrong-Jones in 1960 and the couple had two children, David and Sarah, before they were divorced.

13" composition Alexander Princess Elizabeth, circa 1937. She has sleep eyes, closed mouth, a human hair wig, and is fully jointed. This doll is in mint condition and still has her box, tag, and purse, and is all original. Her tag reads "Authentic/Princess/Elizabeth/Created by/Madame/Alexander." The doll has the Betty face and this model is much more rare than the open mouth Princess Elizabeth dolls (MIB $900+). *Doll and photograph from the collection of Veronica Jochens.*

Several Princess Elizabeth dolls were pictured in the John Plain catalog for 1940. These dolls appear to have open mouths with teeth. Sizes listed for the dolls were 13-1/2", 18", 20", and 24". All had real hair wigs. A small 13-1/2" doll was included with a trunk full of clothes. Included were a play dress, afternoon tea dress, party dress, formal gown, cape, playsuit, housecoat, straw hat, extra shoes, socks, and handkerchiefs. *From the collection of Marge Meisinger.*

Left:
13" composition Alexander Princess Elizabeth dolls, circa 1938-41. They have sleep eyes, open mouths with teeth, mohair wigs, and are fully jointed. These dolls came in a variety of sizes and most of them were marked "Princess Elizabeth/Alexander Doll Co." on the heads. The dress tags read "Madame Alexander New York USA." The two dolls are all original and still have their original boxes. They belonged to two sisters and have been together all their lives (MIB $800+ each). *Dolls and photograph from the collection of Veronica Jochens.*

Right:
18" hard plastic Alexander Margaret Rose, circa 1949. She has sleep eyes, closed mouth, mohair wig, and is fully jointed. She is all original wearing a pink gown tagged "Margaret Rose." She also has her original Alexander wrist tag reading "A/Madame/Alexander/Doll." ($800+). *Doll and photograph from the collection of Nancy Roeder..*

Juanita Quigley

Juanita Quigley was a child actress who played roles in the movies, under the stage name Baby Jane, from 1933 through 1936. She continued in films under her real name from 1936 until 1944. She ended her career at the age of eighteen and entered a convent several years later where she became a nun. After thirteen years she left the convent, went to college, married, and became a teacher.

Her films included *The Great Ziegfeld*, *Born to Dance*, *National Velvet*, *A Yank at Eton*, and *The Blue Bird*.

A composition Baby Jane doll was marketed by the Madame Alexander company to represent the young actress in 1935. The doll was 17" tall, had sleep eyes, open-closed mouth, mohair wig, and was fully jointed. Her head was marked "BABYJANE/REG/MME. ALEXANDER."

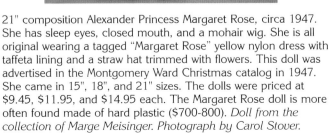

21" composition Alexander Princess Margaret Rose, circa 1947. She has sleep eyes, closed mouth, and a mohair wig. She is all original wearing a tagged "Margaret Rose" yellow nylon dress with taffeta lining and a straw hat trimmed with flowers. This doll was advertised in the Montgomery Ward Christmas catalog in 1947. She came in 15", 18", and 21" sizes. The dolls were priced at $9.45, $11.95, and $14.95 each. The Margaret Rose doll is more often found made of hard plastic ($700-800). *Doll from the collection of Marge Meisinger. Photograph by Carol Stover.*

16" composition Alexander Baby Jane, circa 1935. She has sleep eyes, open mouth with teeth, and a mohair wig. She is marked on the head "Baby-Jane/Reg./MME Alexander." She wears a tagged original red pique dress with patent shoes. The doll was made in the image of the child movie star whose real name was Juanita Quigley ($800+). *Doll from the collection of Marge Meisinger. Photograph by Carol Stover.*

Sabu

Sabu Dastagir was born in India in 1924 and had parts in several movies during the 1940s. He is best known for his role in Alexander Korda's *The Thief of Bagdad* in 1940.

He was marketed in doll form by Molly'es circa 1940 when the firm costumed several dolls to represent characters from the Korda film. The composition Sabu doll was 15" tall, with painted features, and was fully jointed. His costume included a turban which covered his hair. He came with a tag which stated he was Sabu and that he was inspired by Alexander Korda's *Thief of Bagdad*.

Sabu's other films include *Arabian Nights*, *The Jungle Book*, *White Savage*, and *Tangier*. He died of a heart attack in 1963.

Barbara Ann Scott

Barbara Ann Scott became the winner of the Olympic gold medal in figure skating in 1947. She represented Canada in the competition. Because the people in her country were so proud of her accomplishment, the Reliable Toy Co. marketed a doll in her image in 1949. The composition doll was 15" tall and had sleep eyes, open mouth with teeth, a mohair wig, and was fully jointed. The head of the doll was marked "Reliable Doll/Made in Canada." She came with an ice skate shaped tag identifying her as Barbara Ann Scott. The dolls were dressed in ice skating costumes. She is a hard celebrity doll to locate and is very collectible.

Anne Shirley

Anne Shirley was born Dawn Evelyeen Paris in New York in 1918. She began her career as a child model. As she progressed to movies, she used the name Dawn O'Day in her early pictures. In 1934, she played the lead in the RKO film *Anne of Green Gables* and she took the name of the lead character, Anne Shirley, as her own. She acted in many more pictures before she retired at the age of twenty-six. Her last film was *Murder, My Sweet* in 1945.

The actress was married to movie star John Payne from 1937 until 1943. After a divorce, she later married screen writer Charles Lederer and lived a quiet family life with her husband.

The first Effanbee Anne Shirley dolls were marketed circa 1934-35 to tie in to the *Anne of Green Gables* film. The doll company used dolls they were already making and dressed them in Anne Shirley costumes. These included Patsy-Joan, Patricia, Patsyette, and Patsy Jr. dolls.

Later in the 1930s, a new doll was designed which carried the Anne Shirley name on its back. This composition doll came in a variety of sizes. These same doll bodies and sometimes heads were also used on other dolls, including the Little Lady series, many of the American Children dolls, and other dolls wearing costumes not associated with the Anne Shirley character. The dolls that truly represented that character came with braids.

The John Plain catalog for 1940 advertised an Effanbee 15" Anne Shirley doll. The copy reads that the doll was "designed by one of America's foremost child artists." It must refer to Dewees Cochran, as Anne Shirley dolls were available as part of the American Children line. *From the collection of Marge Meisinger.*

17" composition Effanbee Anne Shirley, circa 1939-40. She has sleep eyes, open mouth, a human hair wig, and is fully jointed. She is all original and includes both her metal heart Effanbee bracelet and her original wrist tag. Her tag reads "I am one of/AMERICA'S CHILDREN/Anne/Shirley/AN EFFANBEE/PLAY PRODUCT." The doll was designed by Dewees Cochran and she is marked "Anne Shirley" on her back ($800-900). *Doll from the collection of Lois Jakubowski. Photograph by Robert Jakubowski.*

Jane Withers

Jane Withers was born in Atlanta, Georgia in 1926 and was already an experienced radio performer when she reached Hollywood five years later. Her Atlanta weekly radio program was called "Aunt Sally's Kiddy Club" and the children sang, did imitations, and acted in skits to entertain the radio audience.

Once in Hollywood, Jane received a small part in a film called *Handle With Care* in 1933, but outside of a few extra calls, her career was soon stalled. She turned to radio again, where she became successful on a children's radio show in Los Angeles. When Fox was looking for a child to play opposite their darling Shirley Temple in *Bright Eyes*, someone remembered the little girl from the radio program and Jane was picked for the part. The public loved Jane's performance in the film and wanted to see more of her.

Jane began making films as fast as she could, developing her tomboy character in movie after movie. She starred in *Ginger, This is the Life, Paddy O'Day, Gently Julie, Little Miss Nobody, Pepper, The Farmer Takes a Wife*, and *Can This Be Dixie?* and she was still only nine years old.

Because of Jane's success as a child actress, the Madame Alexander Co. produced a Jane Withers doll in her image in 1937. The dolls came in either closed mouth or open mouth models in sizes of 13", 15", 17", or 20". They were marketed until 1939. Although the dresses carried the Alexander tag, many of the dolls were not marked.

By the 1940s, most of the Withers films had deteriorated to "B" type pictures, like *A Very Young Lady* (Twentieth Century-Fox), *The North Star* (RKO), *Golden Hoofs* (Fox), and *Faces in the Fog* (Republic).

In 1947, Withers gave up movies for marriage, but she did return to Hollywood in 1956 when she played a very good character role in the blockbuster film, *Giant*, made by Warner Brothers.

Jane Withers later turned to television commercials and found her most lasting fame when she played the plumber Josephine. In this role she promoted Comet Cleanser for seventeen years.

17" composition Alexander Jane Withers, circa 1937. This is the same doll advertised in the Sears 1937 catalog. She has a mohair wig and wears a pink organdy dress and a straw hat ($1,200-1,400). *Doll from the collection of Marge Meisinger. Photograph by Carol Stover.*

The Alexander Jane Withers' dolls were featured in the Sears catalog for 1937. The dolls came in sizes of 13-1/2", 15", 17", and 20". They were priced at $2.25, $2.69, $3.59, and $4.59 each. Each doll came with a "Jane Withers" pin and a mohair wig. The smallest size had a closed mouth while the other dolls had open mouths with teeth. *From the collection of Marge Meisinger.*

13" composition Alexander Jane Withers, circa 1937-39. This is the small doll with a closed mouth. She is all original with her pin ($1,000+). *Doll from the collection of Lois Jakubowski. Photograph by Robert Jakubowski.*

178 Personality Dolls

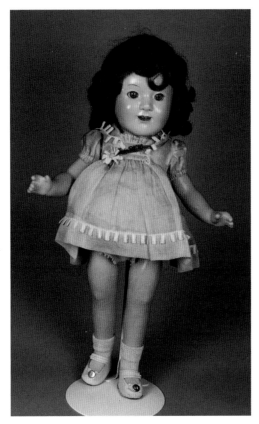

16" composition Alexander Jane Withers, circa 1937-39. She has a mohair wig and an open mouth with teeth. She may have originally had a hat, otherwise she is all original. Some of the dolls were marked "Jane Withers/Alexander Doll" on the back of their heads but many dolls were unmarked except for dress tags or script pins ($1,100+). *Doll from the collection of Marge Meisinger. Photograph by Carol Stover.*

John Plain was still selling Jane Withers dolls in 1939 but their popularity had diminished. The dolls were sold in only two sizes, 13" and 20", in this catalog. *From the collection of Marge Meisinger.*

16" composition Alexander Jane Withers, circa 1937-39. She is wearing her original pink jumper-type dress, black shoes and socks, and hat. Although she has a Jane Withers pin, it may not be original to the doll. The dolls usually came with a pin that had her name spelled out in gold script ($1,150+). *Doll from the collection of Lois Jakubowski. Photograph by Robert Jakubowski.*

Patterns, Doll Clothing, and Mannequins

The Depression years of the 1930s probably produced more home sewn doll clothes than any other era. Every family had a rag bag, so material for doll clothes was close at hand. Old dolls could be given new outfits and a new life more cheaply than it would cost to buy another doll. In this way, dolls were handed down to younger children whenever possible. The dime stores stocked shoes and wigs for dolls at reasonable prices and a "new" dress sewn from material out of the rag bag made the doll "good as new"—or so the mother hoped when this was the main present for a child at Christmas.

Some children were more fortunate and did receive a new doll as a gift, and in the early 1930s, many stores and catalogs featured inexpensive Patsy type composition dolls complete with trunks and wardrobes. The Nancy and Kewty dolls sold by Arranbee were packaged in this manner (see Arranbee chapter). The clothes provided for these dolls were often cheaply made but usually included a "good dress," some type of play wear, pajamas, and a coat and hat.

As the "drink & wet" dolls became more popular, these dolls were also sold with layettes ranging from very simple to quite elaborate, depending on the price of the set.

Most little girls still depended on homemade doll clothes as the Depression years droned on. The commercial pattern companies of the 1930s, who produced patterns for doll clothes, were heavily influenced by the popularity of the Patsy and Shirley Temple type dolls. Many of their patterns were geared to this market. Beginning as early as 1931, McCall Corporation issued patterns for the Patsy family of dolls. Butterick and Simplicity soon followed with their own similar patterns.

With the arrival of the Ideal Shirley Temple dolls in 1934, companies began marketing patterns for clothing to fit these new dolls. Apparently to avoid paying any royalties, the patterns stated they were for the popular "Movie Dolls" or "Film Star Dolls." McCall, Simplicity, Du Barry, Pictorial Review, and Hollywood Patterns all issued patterns of this type. These doll dresses, made at home, reflected the styles made so famous by Shirley in her movies.

In the early to mid 1940s, probably the two most popular doll clothing patterns were those issued by McCall for Effanbee's Dy-Dee Babies and their Little Lady dolls (see Effanbee chapter). The Dy-Dee pattern was copyrighted in 1938 and was still being sold in 1949. The Little Lady pattern carried a copyright of 1943 and was still being advertised by McCall in their *McCall Needlework* magazine in 1949.

Besides patterns for clothing for these popular dolls, the companies also supplied patterns for dolls of all types, including babies, toddlers, and large and small little girl dolls.

The 1940s brought more prosperity and many mothers used new material to make doll dresses that matched a child's dress so they could "look alike." Such a dress was made using the same material and in the same style as the little girl's dress was fashioned. The red, white, and blue colors found in many of these outfits reflected a time period when patriotism was high due to World War II.

Dresses with matching underwear were also made in the early 1940s. This type of one-piece underwear was popular for dolls from the 1920s through the 1940s. It was cheaper and easier to make just one piece instead of both a slip and panties. If a slip was made, it usually included attached panties instead of being made separately.

By the 1940s, wardrobes for dolls could be purchased through catalogs and from local dime stores. These clothing items were sold very cheaply and as mothers began to be involved in jobs during World War II, their time for making doll clothes was limited. With the end of the Depression, families had more money to spend and these commercially made wardrobes seemed to sell well.

The composition mannequin sewing dolls became popular in the 1940s and several different brands were made, including dolls by Simplicity, McCall, and Singer as well as the Junior Miss doll (which included patterns by Butterick).

The Simplicity dolls could be purchased in a variety of kits. The dolls came in two sizes, 12-1/2" and 15", and were sold with or without a dress form. Three patterns were included in each kit. The prices ranged from $1.79 to $3.50 for each set.

Like the Simplicity doll, the Peggy McCall fashion doll was all composition with removable arms. This nearly 13" doll also came in a boxed set. It included a dress form, three patterns, and accessories. Like most of these sewing sets from the World War II era, a pattern for a nurse's uniform was provided.

The 14" Junior Miss mannequin from Educational Crafts Co. came supplied with Butterick patterns to make a bridal dress, formal, "dress-up" dress, and W.A.A.C, W.A.V.E, and nurse's uniforms.

Another less seen mannequin set was marketed under the name, "Susanne's Fashion Show." It carried the copyright of Latexture Products of New York City. This mannequin also came with three patterns, including those for a dress, nurse's uniform, and nurse's cape.

Besides these mannequin type sewing kits, other cheaper sewing sets were also being marketed. These usually included an all composition little girl doll varying in size from 6" to 9-1/2" with very simple cut-out pieces of cloth to be sewed up the sides to fit the dolls. "Miss Deb Sewing Kit" and "Jean Darling Doll Sewing Kit" are two examples of this type of set.

By the late 1940s, the mannequin dolls had lost favor but the Singer Manufacturing Co. offered a new example in 1949. Instead of being made of a composition like material, this new example was manufactured of soft vinyl. The head appears to be too small for its body. The doll came with two Butterick patterns to make various dresses, a skirt, blouse, jacket, and a pinafore.

Nealy all of the patterns that came with the mannequin dolls were very difficult to sew and when the garments were finished, they were often too tight to fit the doll properly. A lot of these kits were sold before consumers lost interest in these "sewing for dolls" enterprises.

By the 1940s, home dressmakers were becoming much more skillful in making doll clothes. Perhaps the intricate patterns from the mannequin dolls influenced the other commercial doll patterns as they, too, became more complicated.

As hard plastic dolls took over the market in the late 1940s, dolls were provided with more glamorous clothes and homemade doll clothes followed the trend with fancy evening clothes, bridal dresses, and lounging attire. With the coming of the 1950s and the many commercial outfits provided for the Ginny, Ginger, Muffie, and other similar dolls, homemade doll clothing became an even smaller part of the doll world.

About the only company that was still publicizing home sewing for dolls in the 1950s was the Mary Hoyer firm. Books were sold with detailed patterns to make doll clothes either to sew, knit, or crochet for the Hoyer dolls (see Mary Hoyer chapter).

Those collectors who remember special handmade doll outfits lovingly provided by a mother or grandmother still treasure the "made at home" clothing now in their collections. It is one way to remember, as this author does, a summer afternoon spent sitting on a blanket in the backyard as a mother sewed a new garment for a favorite doll to add to the trunk of clothes already made. The excitement of a little girl as she waits to try the completed dress on her favorite Wendy Ann doll can still be remembered with pleasure.

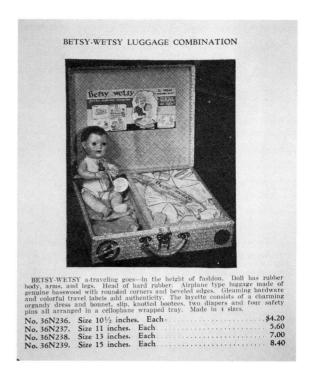

The Ideal Betsy-Wetsy drink & wet doll came in sizes of 10-1/2", 11", 13", and 15" in this advertisement from the N. Shure Co. catalog for 1938. A layette was also provided for each doll. The doll had a hard rubber head and a soft rubber body. *From the collection of Marge Meisinger.*

Dolls and clothing packaged in trunk sets were featured in the Montgomery Ward Christmas catalog for 1933. The composition dolls ranged in size from 9-3/4" to 16" tall and were priced from 89 cents to $3.98 for each set. The dolls and clothing all had the Effanbee Patsy look but the dolls were made by lesser known companies. *From the collection of Marge Meisinger.*

The McCall Co. was one of the first to market commercial doll clothes patterns to fit the Effanbee Patsy dolls. This #1919 was made to fit the 11-1/2" Patsykins doll and carries a copyright of 1931. These patterns can still be used to dress dolls in the proper clothing of the period ($40+). *From the collection of Marge Meisinger.*

Patterns, Doll Clothing, and Mannequins 181

McCall pattern from the same era made to fit a 22" Patsy type doll. It was copyright 1933 ($25+).

Butterick Pattern #442 was made to fit a 20" Patsy type doll. The pattern included two dresses, hat, beret, coat, pajamas, bathing suit, and combination slip and "step-in" ($35+). *From the collection of Marge Meisinger.*

This 19" Effanbee composition Patsy Ann wears a dress, one piece underwear, and hat made by a devoted grandmother in the 1930s. The material is pongee. The outfit is similar to those shown in the patterns offered by McCall.

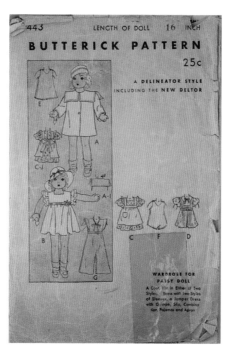

Butterick Pattern #443 was made to fit a 16" Patsy family doll. It included a coat and hat, dress and hat, jumper dress, slip-panties combination, apron, and pajamas ($35+). *From the collection of Marge Meisinger.*

Simplicity pattern #1901 features clothes with the Patsy look even though it doesn't say the pattern is for the Patsy dolls. This one was made to fit a 20" doll ($30+). *From the collection of Marge Meisinger.*

McCall pattern #355, copyright 1935, made to fit the "Popular Movie Dolls" (Shirley Temple and look-a-likes). It is the 18" size ($35+). *From the collection of Marge Meisinger.*

McCall pattern #418 made to fit an 18" Movie Doll. The pattern carried a copyright of 1936. The drawings on the front of the pattern look so much like the Shirley Temple dolls, there is no doubt who the "Movie Doll" was intended to be ($40+). *From the collection of Marge Meisinger.*

Simplicity Pattern #2240 made to fit a 24" doll who looks similar to Shirley Temple on the cover, although the star is not mentioned ($35+). *From the collection of Marge Meisinger.*

Du Barry pattern #1616B made to fit an 18" Film Star Doll. The pattern includes a coat and hat, pajamas, dress, and sunsuit ($30+). *From the collection of Marge Meisinger.*

Patterns, Doll Clothing, and Mannequins 183

McCall pattern #525 made to fit a 13" Movie Doll. The drawings on the front of the pattern indicate the clothes were for a Shirley Temple doll ($40+). *From the collection of Marge Meisinger.*

Hollywood Pattern of Youth #1913 from 1942 was also geared to the Film Star Dolls. This one was for the 20" size dolls ($30+). *From the collection of Marge Meisinger.*

Pictorial Review pattern #8507. The printing on the front says that the clothes will fit the popular Film Star Dolls 18" in size ($30+). *From the collection of Marge Meisinger.*

Directions for a Shirley Temple type dress that could be made for a little girl and her doll. The doll dress is pictured. The pattern was issued by the American Thread Co. of New York and the leaflet was No. 45, 11-35. The pattern was for a 13" doll. There is no mention of Shirley in the printed material (dress and pattern $50+). *From the collection of Marge Meisinger.*

13" Ideal composition Shirley Temple is pictured with wardrobe pieces made by a grandmother for her granddaughter, circa 1936-38. Included is the dress she is wearing which is trimmed with commercial ruffling, a taffeta dress and matching one piece underwear, and a white pongee coat lined with red ($20-35 each outfit, clothing only).

McCall pattern #1098 for clothes to fit the Effanbee Little Lady dolls, size 18". Pattern includes a negligee, suit and hat, bridal gown, slip and panties. The pattern carries a copyright of 1943 ($30+).

McCall pattern #632 was for clothes to fit the Effanbee Dy-Dee Dolls, size 20". It carries a 1938 copyright. Pattern includes a coat and bonnet, romper, sunbonnet, dress and bonnet, slip, and bunting ($30+).

McCall pattern issued in 1937 for little girl dolls 22" tall. There is no reference to any particular "named" doll. The package includes patterns for a dress, hat, and a slip with attached panties ($25+). *From the collection of Marge Meisinger.*

Patterns, Doll Clothing, and Mannequins 185

Right:
Du Barry Pattern #2144B to fit a 16" little girl doll, circa early 1940s. A coat and hat, made from the pattern, is modeled by a 16" unmarked composition doll with painted eyes, closed mouth, and a mohair wig. She wears a cotton printed dress made from the pattern under her coat. Clothes were made by the author's mother, Flossie Scofield, in the 1970s (pattern $25+, coat, hat, and dress $40+).

Advance pattern #2616 to fit a 16" little girl doll, circa 1940s. Included were patterns to make a coat and hat, pinafore and dress, dress with a jumper look, robe, nightgown, and formal. (30+). *From the collection of Marge Meisinger.*

Simplicity Pattern #4851 made to fit a 12" little girl doll, circa 1940s. This pattern was very popular and was sold for many years. ($30+). *From the collection of Marge Meisinger.*

18" unidentified composition doll with sleep eyes and human hair wig. She is pictured with some of the items in her wardrobe from the mid 1940s. Her original wig is also in the photo. Even the skating outfit she wears was made by the owner's mother. *Childhood doll of Marilyn Pittman. Photograph by Marilyn Pittman.*

186 Patterns, Doll Clothing, and Mannequins

12" composition Kewpie dolls, circa 1947. They are dressed in handmade clothing styled to match outfits worn by two young sisters. *Childhood dolls of Bonnie McCullough and Marilyn Pittman. Photograph by Marilyn Pittman.*

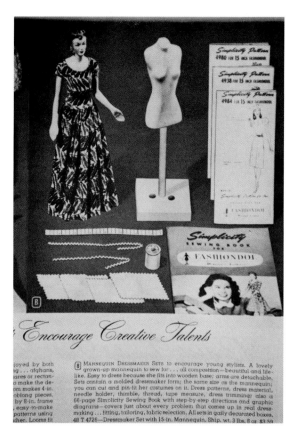

The Montgomery Ward Christmas catalog for 1943 featured the Simplicity mannequin set in two sizes with either a 12-1/2" or a 15" doll. The sets sold for $1.79 and $3.59 each. The more expensive set also included a dress form. The sets came with patterns and accessories.

Three different sets of doll clothing were advertised in the Montgomery Ward Christmas catalog for 1940. The cheapest set included two dresses, two bonnets, slip, panties, and hangers for just 53 cents. *From the collection of Marge Meisinger.*

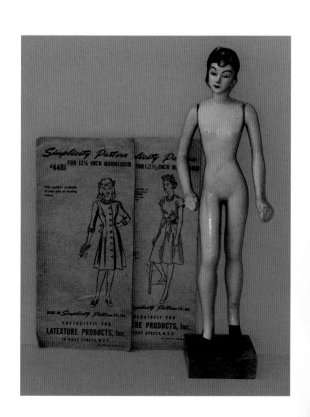

Right:
12-1/2" Simplicity mannequin, circa 1943. The doll has painted features and molded hair. Like all mannequin dolls of the period, this doll's arms are removable, which made it easier to fit the clothing. The dolls came with patterns and other sewing accessories. The two dress patterns shown with this doll are marked "Made by Simplicity Pattern Co., Inc. Exclusively for Latexture Products, Inc. N.Y.C." Although the Montgomery Ward catalog said the doll was made of composition, it may have been made of Latexture. A stand with two holes came with the doll to allow it to stand upright ($100+).

12-1/2" Peggy McCall composition mannequin doll, circa 1942. She has painted features, molded hair, and removable arms. The patterns which accompanied the doll are labeled "McCall Printed Pattern. Copyright 1942 by McCall Corporation." The three patterns include a formal, dress, and nurse's outfit. A dress form is missing from the set. These doll packages were advertised by the Dritz-Traum Co., of New York City, in 1943. This company must have been involved in the enterprise, perhaps being responsible for the manufacture of the dolls. This is one of the most desirable of the mannequin dolls because it is harder to find ($200+ with crazing).

13" composition Junior Miss mannequin doll manufactured by Educational Crafts Co., New York City. The doll has painted features, molded hair, and removable arms. She came with patterns supplied by Butterick. The costumes include a dress, formal, wedding gown, and W.A.A.C, W.A.V.E, and nurse's uniforms. Two additional patterns could be purchased for 25 cents each, which included pajamas, slip, robe, evening gown, and princess frock. The doll is pictured in her wedding gown ($135+ with repaired arm).

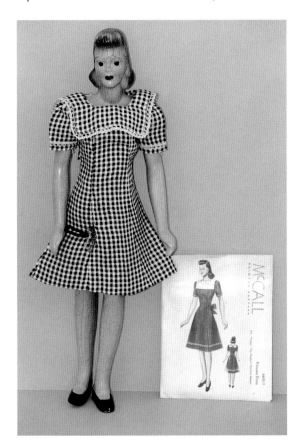

The Peggy McCall doll models a dress made from one of the patterns included in the set.

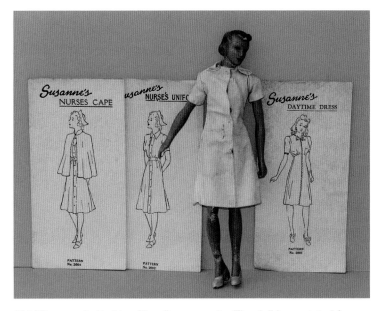

11" "Susanne's Fashion Show" mannequin. The doll has painted features, molded hair, and removable arms. According to printed material accompanying the doll, she was made of a flexible material. With the passage of time, however, she is no longer flexible. Her copyright was listed as being held by Latexture Products, Inc. New York City. Unlike the other mannequin dolls, the clothes patterns were not made by a famous pattern designer. The patterns include a dress, nurse's dress, and a nurse's cape. The doll is dressed in her nurse dress made from the original pattern. This mannequin is not easy for a collector to find ($75 plus with mended ankle).

188 Patterns, Doll Clothing, and Mannequins

"Miss Deb Sewing Kit" made by the Pressman Toy Corp. of New York, New York, circa late 1940s. The kit includes a 9" composition doll with painted features, molded hair, and jointed arms. Also provided are six simple pre-cut outfits, three felt hats, tape measure, thread, simple scissors, and needles. Many different companies made these cheaply priced sewing kits during the 1930s and 1940s. The earlier kits included painted over bisque or white bisque dolls instead of composition. This set is unusual because the doll is so large. The dolls were usually around 5"-6" tall ($30-40).

Hard plastic dolls from the late 1940s model "glamorous" lounging wear from the era. The taffeta robe is a commercial product while the quilted jacket and taffeta pants were handmade. The doll on the right is a 14" Ideal Toni while the hard plastic doll on the left is unidentified (clothing $25-35 each outfit).

12-1/2" "Singer Mannikin Doll Set" from 1949. the doll has painted features, molded hair, and removable arms. She is made of soft vinyl. The copy on the box lid reads "Set made expressly for the SINGER MFG. CO. by W. Smith Industries." The patterns were supplied by the Butterick Co. They included several dresses, a skirt, blouse, jacket, and a pinafore. Additional patterns were also available. The sewing booklet lists a copyright of 1949 ($150+).

Bibliography

Adams, Margaret. *Collectible Dolls and Accessories of the Twenties and Thirties from Sears, Roebuck and Co. Catalogs*. New York: Dover Publications, Inc., 1986.

Alexander Doll Company, Inc. Catalog. New York, 1942.

Anderton, Johana Gast. *Cloth Dolls*. Lombard, Illinois: Wallace-Homestead Book Company, 1984.

———. *Twentieth Century Dolls From Bisque to Vinyl*. North Kansas City, Missouri: The Trojan Press, 1971.

Axe, John. *The Encyclopedia of Celebrity Dolls*. Cumberland, Maryland: Hobby House Press, Inc. 1983.

———. "Interview At the Alexander Doll Factory." *Doll Reader*. November, 1983.

———. "Madame Alexander Receives the Lifetime Achievement Award." *Doll Reader*. May 1986, pp.128-130.

———. "Tribute to the Grand Madame - Madame Beatrice Alexander." *Doll Reader*. December 1990-January 1991, pp. 180-182.

———. "Walt Disney's Pinocchio." *Doll Reader*. February, March 1981, pp. 57-61.

Carlson, Nancy. "Dewees Cochran: Her Life and Legacy." *Doll News*. Winter, 1992, pp. 24-30.

Cobb's Doll Auction Catalogs. Johnstown, Ohio. From 1988-2000.

Cochran, Dewees. *As If They Might Speak*. Santa Cruz, California: Paperweight Press, 1979.

Coleman, Dorothy S. and Evelyn Jane Coleman. "Joseph L. Kallus." *Doll Reader*. November 1989, pp. 156-161.

Comienski, Barbara Lutz. "Effanbee's Little Lady - Anne Shirley." *Doll Reader*. May 1988, pp. 142-145.

Foulke, Jan. *Blue Book Dolls and Values*. Volumes Three-Fourteen. Cumberland, Maryland: Hobby House Press, Inc.

———. "The Million Dollar Baby." *Doll Reader*. October 1992, pp. 114-121.

———. "Raggedy Ann Sweet All The Way Through." *Doll Reader*. September 1992, pp. 82-89.

Frasher's Doll Auction Catalogs. Oak Grove Missouri. From 1998-2000.

Hays, Ann. "The Many Faces of Georgene Averill." *Doll Reader*. May 1984, pp. 72-73.

Hoyer, Mary. *Mary Hoyer and Her Dolls*. Cumberland, Maryland: Hobby House Press, Inc. 1982.

Hunter, Sheryl. "She's Sassy She's Flasy...She's Betty Boop." *Doll World*. June 1988, pp. 26-28, 52-53.

Izen, Judith. *Collector's Guide to Ideal Dolls*. Paducah, Kentucky: Collector Books, 1999.

———. "Make-up Dolls by Vogue." *Doll Reader*. January 1999, p. 150.

Izen, Judith, and Carol Stover. *Collector's Encyclopedia of Vogue Dolls*. Paducah, Kentucky: Collector Books, 1998.

John Plain Catalogs. Chicago, Illinois. 1937, 1938, 1939, 1940.

Judd, Pam, and Polly Judd. *Compo Dolls 1928-1955*. Cumberland, Maryland: Hobby House Press, 1991.

———. *Hard Plastic Dolls Identification and Price Guide*. Cumberland, Maryland: Hobby House Press, 1990.

———. "The Many Versions of Snow White and the Seven Dwarfs." *Doll Reader*. June-July 1988, pp. 126-131.

———. "World War II Dolls in Uniform." *Doll Reader*. October 1986, pp. 116-119.

Judd, Polly. *Cloth Dolls Identification and Price Guide*. Cumberland, Maryland: Hobby House Press, 1990.

Larkin Co. Catalogs. Buffalo, New York. 1935, 1937.

Lavitt, Wendy. "Dewees Cochran 1892-." *The Collector's Magazine*. September-October 1987, pp. 48-51.

Leeser, Paulus (Photographer). "Portrait Dolls." *Pageant*. December 1945, p. 81.

Majors, Louise Fultz. "The Kewpie Doll Kraze." *Doll World*. August 1988, pp. 60-62.

Mary Hoyer Doll Mfg. Co. Reading, Pa. Various company booklets and brochures.

McCall Corporation. *McCall Needlework*. Winter, 1949-50, p.86. Fall-Winter 1950-51. p.93.

Meisinger, Marge. "Gone With The Wind - Movie and Dolls." *Doll Reader*. January 1990, pp. 192-199.

Mertz, Ursula R. "American Doll Showcase - Kiddie Pal Dollies." *Doll Reader*. January 1999, p. 24.

———. ""American Doll Showcase - Positively Identifying Arranbee Dolls." *Doll Reader*. May 1999, pp. 20-22.

———. *Collector's Encyclopedia of American Composition Dolls 1900-1950*. Paducah, Kentucky: Collector Books, 1999.

———. "The Composition Campbell Kids Dolls and Their Look-alikes." *Doll Reader*. May 1989, pp. 100-104.

Miller, Marjorie A. *Nancy Ann Storybook Dolls*. Cumberland, Maryland: Hobby House Press 1991.

Montgomery Ward Catalogs, Chicago, Illinois. Various issues.

N. Shure Co. Catalogs, Chicago, Illinois. 1938, 1941-42.

Nagley, Richard Ashton, and Tracy Sexton Davis. "Little Lulu, I Love You, Lu." *Doll Reader*. December 1988, January 1990, p. 177.

O'Sullivan, Betty. "From the 'As If They Might Speak' Exhibit Dewees' Darling Dolls." *Doll News*. Fall 1993, pp. 58-61.

Pardella, Edward. "Baby Sandy Dolls." *Doll Reader*. July 1992, pp. 114-118.

Ratcliff, Mary Lou. "Fabulous Maud Tousey Fangel" in *Best of Doll Reader Vol. IV*. pp. 140-145. Cumberland, Maryland: Hobby House Press, 1991.

Reed, Sonja. "Dewees Cochran, First Among Equals." *Doll World*. February 1997, pp. 28-30.

Robison, Joleen Ashman, and Kay Sellers. *Advertising Dolls*. Paducah, Kentucky: Collector Books, 1992.

Schmuhl, Marian H. "Ralph Freundlich." *Doll Reader*. February 1994.

Schoonmaker, Patricia N. "Composition Corner." *Doll Reader*. May 1984, p. 117. May 1987, p. 157. May 1988, p. 171. February-March, 1991, p. 131. July 1999, p. 108.

———. *Patsy Doll Family Encyclopedia Volume I*. Cumberland, Maryland: Hobby House Press, 1992.

Sears Catalogs. Chicago. Various issues.

Sieverling, Helen. "The Many Faces of Santa Claus." *Doll Reader*. December 1988-January 1989, pp. 81-83.

Spiegel Catalogs. Chicago. 1939, 1941.

Smith, Patricia. *Doll Values Antique to Modern*. Volumes Two-Twelve. Paducah, Kentucky: Collectors Books.

———. *Effanbee Dolls That Touch Your Heart*. Paducah, Kentucky: Collector Books, 1983.

———. *Madame Alexander Collector's Dolls*. Paducah, Kentucky: Collector Books, 1977.

———. *Madame Alexander Collector's Dolls Price Guide*. Paducah, Kentucky: Collector Books, 1996.

Stuecher, Mary Rickert. "Drink and Wet Babies The Dy-Dee Doll" Part I. *Doll Reader*. December 1983-January 1984, pp. 90-98.

Tardie, Ann. "Creations in Cloth by Madame Alexander." *Doll Reader*. November 1986, pp. 126-131.

Theriault's Doll Auction Catalogs. Annapolis, Maryland. 1999-2000.

Young, Mary. *Tomart's Price Guide to Lowe and Whitman Paper Dolls*. Dayton, Ohio: Tomart Publications, 1993.

Walker, Z. Frances. "Hedwig Dolls." *Doll Reader*. August-September 1983, pp. 138-139.

———. "A Maud Tousey Fangel Model - Mabelle Lynch." *Doll Reader*. October 1989, p. 206-210.

Zillner, Dian. *Collectible Coloring Books*. West Chester, Pennsylvania: Schiffer Publishing Ltd., 1992.

———. *Hollywood Collectibles*. West Chester, Pennsylvania: Schiffer Publishing Ltd., 1991.

———. *Hollywood Collectibles The Sequel*. Atglen, Pennsylvania: Schiffer Publishing Ltd., 1994.

Mickey & Janie, 48
Miller, Marjorie A., 99
Mimi, 46
Miranda, Carmen, 172-173
Miss Curity, 92
Miss Deb Sewing Kit, 188
Miss Ginger, 85-86
Mollye/International Doll Co., 136, 137
    Russia, 137
Molly'es
    Raggedy Andy, 131
    Sabu, 176
Monica Doll Studios, 97-98
Mortimer Snerd, 85, 161, 171-172
Mother & Me, 9, 20-21
Mountie, 118
My Betty, 6

N

Nancy, 31-33, 36-37
Nancy Ann Storybook Dolls, 99-105
    Around the World-Scotch, 100
    August, 105
    Ballet Dancer, 102
    Blackeyed Susan, 102
    Bo Peep, 100, 101
    Boudoir Box, 103
    Bride, 105
    Chinese, 102
    Cowboy, 102
    December, 104
    Gypsy, 102
    Hansel & Grethel (sic), 101
    He Loves Me, 101
    Hush-a-Bye Baby, 99, 100
    I have a Little Pet, 101
    Judy Ann, 99, 101
    Lady in Waiting, Charmaine, 104
    Mammy, 101
    Margie Ann, 103
    Mistress Mary, 104
    Naughty Marietta, 105
    Pirate, 102
    Pussy Cat, 103
    Queen of Hearts, 105
    Riding, 102
    Skiing, 102
    To Market, 101
    Topsy & Eva, 103
Nancy (comic), 73
Nancy Lee, 31, 34-36
Nancy the Movie Queen, 33
Naughty Sue, 77
Nun Doll, 142
Nurse, 145-147

O

O'Brien, Margaret, 9, 23-24, 173-174
O'Neill, Rose, 129-130
    Giggles, 129
    Kewpie, 129, 138, 186
    Scootles, 129

P

Pantalette, 106, 110
Pat-a-Pat, 140
Patricia, 46, 54-55
Patricia Glamour Doll, 63
Patsy, 46, 50, 53-54, 57
Patsy Ann, 46, 55, 57, 181
Patsy Baby, 46, 53
Patsy Babyette, 46, 51-52
Patsy (early), 46, 49
Patsyette, 46, 48, 52-53
Patsy Joan, 46, 55
Patsy Jr.-Patsykins, 46, 53
Patsy Lou, 46, 56
Patsy Mae, 46, 56
Patsy Pattern, 180
Patsy Ruth, 46, 56
Patsy Tinyette, 46, 51
Peggy, 77, 78
Peggy Ann, 73
Peggy Lou, 38, 42
Peggy McCall, 187
Peruggi, Ernesto, 77

Ponsett, 39
Pictorial Review Pattern, 179, 183
Pigtail Sally, 77, 82
Pinkie Alexander, 9, 18
Pinkie (Cameo), 130
Pinocchio, 85, 128, 161
Plassie, 85, 91
Polly, 114
Popeye, 125
Portrait (Effanbee), 48, 66
Portrait Series (Alexander), 8, 10
Prager & Rueben, 148
Pressman Toy Corp., 188
Princess Beatrice, 85, 88
Princess Elizabeth, 8, 12-13, 23, 174-175
Puggy, 26, 30
Putnam, Grace S., 113
Puzzy, 130

Q

Q-T Drinking and Wetting Doll, 121
Queen of the Ice, 85-89
Quintuplets, 31
Quiqley, Juanita, 175

R

Raggedy Ann and Andy, 73, 76, 131
Raggy Doodle U.S. Parachute Jumper, 148
Red Cross Dolls, 147
Reliable Toy Co. 118, 145, 155, 176
    Barbara Ann Scott, 176
    Her Highness, 118
    Indian, 118
    Mountie, 118
    Nurse, 145
    Shirley Temple, 155
Regal Doll Manufacturing Co., 77, 117
    Judy Girl, 117
    Kiddie Pal, 117
Rita Hayworth, 121, 165
Roberta, 77, 79
Roberta Doll Co., 119
    Bride, 119
Rosalie, 31, 35
Rosebud, 77
Rowland, Les, 99
Rumbera and Rumbero, 10

S

Sabu, 176
Sailor, 63, 144, 146-148
Sally, 26, 29
Sally Jane, 26, 29
Sandra Pla-mate, 72
Santa Claus, 132
Scarlett O'Hara, 8, 169-170
Schmuhl, Marian, 114
Scott, Barbara, Ann, 176
Scootles, 129
Segar, Elzie Crisler, 125
Shirley, Anne, 176
Shirley Temple, 85, 149-155, 162, 179, 184
Simplicity Doll, 186
Simplicity Patterns, 46, 179, 181-182, 185-186
Singer Mannikin Doll, 188
Sizzy, 130
Skippy, 46, 63, 125, 144
Sleeping Beauty, 24
Sleepy-time Twins, 85, 91
Sluggy, 73
Slumbermate, 9
Smith, Ira, 10
Snoozie, 85-86
Snow White, 8, 38, 85, 126-128
So-Lite Baby, 134
Soldier, 63, 144-148
Sonja Henie, 8, 164, 166-168
Sonja on Skates, 38, 47, 61
Southern Girl, 8, 22
Southern Series 31, 33, 36
Spanish Dancers, 48, 66
Special Girl, 8, 12
Stover, Carol, 106
Su'ben's Art, 39
Sun Rubber Co., 119-120, 160
    Amosandra, 160

    Drink & Wet Dolls, 119-120
Sunshine Baby, 106-108
Superman, 126
Susan Stormalong, 39
Susanne's Fashion Show, 187
Susie-Q & Bobby-Q, 133
Suzanne, 48, 66
Suzette, 48, 65
Sweetie Pie, 48, 63, 68-69
Sweetheart, 80

T

Teeny Twinkle, 134
Temple, Shirley, 85, 149-155, 162, 179, 184
Three-In-One Doll Corp., 120
    Three Faced Trudy, 120
Tickle Toes, 85, 88
Tinie Baby, 79
Tiny Betty, 6-7, 9
Tiny Town Dolls, 138, 139
    Ballerina, 139
    Dollhouse Family, 139
    Hansel & Gretel, 139
    Red Riding Hood, 138
    Swiss Boy & Girl, 139
Tippietoe, 134
Toddletot, 26, 28
Toddles, 106-108
Tommy Tucker, 48
Toni, 188
Toodles, 26-28
Topsy & Eva, 76
Topsy-Turvy, 11
Tousle Head, 46, 59
Tousle-tots, 52
Trudy (Three face), 120
Tubby, 124
Twelvetrees, Charles, 26, 30

U

Uncle Sam, 145
Uneeda Doll Co., 121
    Rita Hayworth, 165
    Sweetums, 121
    Toddlers, 121

V

Vogue Dolls, Inc., 31, 106-110
    Betty Jane, 106, 108
    Dora Lee, 106, 107
    "Draf-tee," 145
    Joan, 110
    Jennie, 106
    Ginger, 110
    "Just Me," 106
    Make-up Dolls, 106, 108
    Mary Jane, 106, 109
    Pantalette, 106, 110
    Polly, 110
    Sunshine Baby, 106, 108
    Toddles, 106-108, 145
    Uncle Sam, 145
    WAAC-ette, 106
    WAVE-ette, 106
Volland, P. F., 131

W

W.A.A.C., 143, 146
W.A.V.E., 26, 143, 144
W.C. Fields, 163
Wee Patsy, 46, 50
Wee Wee, 26
Well Made Doll Co., 121
    Q-T Drinking & Wetting Doll, 121
Wendy Ann, 8, 16, 17, 22
Winnie Winkle, 126
Withers, Jane, 9, 177, 178
Wondercraft Co., 126
    Denny Dimwit, 126
    Bobbi Mae, 126

Y

Young, Murat (Chic), 124